ANARCHY EVOLUTION

itbooks

AN IMPRINT OF HARPERCOLLINS*PUBLISHERS*

ANARCHY
EVOLUTION

FAITH, SCIENCE, AND BAD RELIGION IN A WORLD WITHOUT GOD

GREG GRAFFIN
& STEVE OLSON

Grateful acknowledgment is made for permission to reprint the following:

"We're Only Going to Die from Our Own Arrogance," music and lyrics by Greg Graffin, from *How Could Hell Be Any Worse* by Bad Religion. Copyright © 1982 by Polypterus Music.

"God's Love," music and lyrics by Greg Graffin and Brett Gurewitz, from *The Empire Strikes First* by Bad Religion. Copyright © 2006 by Polypterus Music and Sick Muse Songs.

"Rip It Up," music and lyrics by Tony Cadena, from *Adolescents* by the Adolescents. Copyright © 1981 by Frontier Records.

"God Song," music and lyrics by Greg Graffin, from *Against the Grain* by Bad Religion. Copyright © 1990 by Polypterus Music.

"Atheist Peace," music and lyrics by Greg Graffin, from *The Empire Strikes First* by Bad Religion. Copyright © 2006 by Polypterus Music and Sick Muse Songs.

HarperCollins books may be purchased for educational, business, or sales promotional use. For information please write: Special Markets Department, HarperCollins Publishers, 10 East 53rd Street, New York, NY 10022.

Designed by William Ruoto

Library of Congress Cataloging-in-Publication Data
Graffin, Greg.
  Anarchy evolution : faith, science, and bad religion in a world without god / Greg Graffin and Steve Olson. — 1st ed.
      p. cm.
  Includes bibliographical references.
  ISBN 978-0-06-182850-8
  1. Religion—Philosophy. 2. Evolution (Biology)—Religious aspects. 3. Religion and science. 4. Atheism. I. Olson, Steve, 1956– II. Title.
  BL51.G7195 2010
  210—dc22

                                                                    2010021739

10 11 12 13 14   OV/RRD   10 9 8 7 6 5 4 3

For
Allison, Graham, and Ella
fcf

For
Sarah and Eric
SO

CONTENTS

ANARCHY EVOLUTION

CHAPTER 1

THE PROBLEM WITH AUTHORITY

Sire, I have no need of that hypothesis.
—Pierre-Simon Laplace[1]

To punish me for my contempt for authority, fate made me an authority myself.
—Albert Einstein[2]

I've always had a problem with authority. When I was in the third grade at Lake Bluff Elementary School just outside Milwaukee, my teacher, Wanda Rood, knew that I hated to be called by my full name, Gregory. I have always been Greg to my family and friends, and whenever Miss Rood called me Gregory to humiliate or intimidate me, I burned with anger. Finally, one day when I was talking too much to my friends, Miss Rood said, "Gregory, do you have something to say to us all?" I replied, "Don't call me Gregory, Wanda."

When my mother heard what I'd done, she laughed. "That mouth of yours is going to get you in trouble," she said. And I did get in trouble, especially after she got back from meeting with the principal. But my mom rarely condemned me or my brother for the rebellious things we did when we were kids.

Maybe she figured that suppressing the bad would get rid of the good, too.

All of us are bombarded every day by people telling us what we should do and how we should think. Politicians try to persuade us that they have the answers to the world's problems if only we will enlist in their cause. Ministers, priests, and imams tell us that we must lead a life in accord with ancient decrees or face the consequences in the afterlife. We are constantly exposed to subtle or not-so-subtle messages about how we should behave—in advertisements, movies, radio and television talk shows, even music and books.

Even more alarming than the number of these injunctions is their frequent intolerance. All my life I have been subjected to the dogmatic, fundamentalist attitudes of authority figures. Maybe I have a congenital distaste for authority, but I feel that I grew up surrounded by fundamentalisms—and today those fundamentalisms seem to be getting even stronger. My family has a history of religious evangelism, but at least my elderly relatives were respectful of others' right to independent thought. Today I can't open a newspaper without reading about religious fanatics detonating bombs that they have strapped to their bodies or harassing abortion providers to the point of death. Political parties demand litmus tests of their members to make sure they will not stray from highly partisan edicts. Even in the fields where I have chosen to spend the majority of my life—music and science— I've often encountered authorities who rarely stray from worn-out dogma and overbearing expectations for allegiance.[3]

There are two possible ways of responding to these demands. One is to acquiesce, whether wholeheartedly or with reservations. Many of my friends are religious, and they've given me many explanations for their beliefs. "Because I want to go to heaven and live forever," they might say. Or "because I want to avoid sin," others respond, or "because I want to live the good life exemplified by the martyrs." These answers are similar to those my nonreligious friends give me when I ask them why they bow to the demands of authority. "Because I don't want to make waves," or "life is easier if we avoid controversy," or "I don't really have my own philosophy, so I might as well try someone else's." People have many ways of justifying their behaviors to themselves and to others.

The other option is to resist authority, whether quietly or defiantly. When Bad Religion released its third album, *Suffer*, in 1988, the album cover showed a teenage boy standing in a bland suburban neighborhood consumed by flames. (The image occurred to me and my friend Jerry, who painted it, while we were working as salad-bar hosts at an L.A. restaurant.) That image seemed to capture the simultaneous anger, powerlessness, and defiance of my teenage years, emotions that fueled my songwriting and performances in the early days of the band. And that image seems also to have resonated with many Bad Religion fans, because I've seen hundreds of torsos and arms emblazoned with "suffer boy" tattoos.

But there is a big problem with resisting authority. If you can't rely on it, what can you rely on? Many people feel very uncomfortable giving up the certainties on which they have based their lives. They think that without a bedrock of belief, their lives will have no purpose or meaning. Many religious people, for example, believe that without religion there can be no morality. They fear that humans might use their free will to do terrible things—steal, rape, murder—if they did not believe in a caring

God constantly watching over and keeping them on the path of righteousness.

For those of us who see no need for supernatural entities, this is a highly offensive belief. It condemns our lives as deviant and amoral. It also has no empirical proof. The countries that are least religious tend to have the most law-abiding and generous citizens.[4] It doesn't even make sense, as philosophers since the time of Socrates have pointed out.[5] Either harming other people is wrong, in which case God is unnecessary, or harming other people is acceptable, in which case God's admonitions are misguided.

Some people have a more basic fear. I've had friends who think that if they question their beliefs in God or spirituality or some "higher purpose," they will begin a long, lonely descent toward nihilistic anarchy in their lives. They fear that they will see themselves as no more than soulless animals, biological mechanisms, bits of a temporary consciousness that will soon be gone forever.

This fear is not entirely misplaced. The natural world and the evolutionary processes that produced us are anarchic. There is no ultimate reason for our existence. We were born to parents who loved us, if we were fortunate, and who wanted us to do well in life. But we were not placed on this earth for some divine purpose that only communion with the spirit world can reveal.

However, people make a big mistake if they conclude from the anarchy of the physical world that life has no meaning. I draw just the opposite conclusion. The purposelessness of the natural world emphasizes the tremendous meaning inherent in the human world.

It took me a long time to figure out and be able to describe what I rely on in place of authority. I had to experiment with different ideas. I had to get into the world to see which ideas worked and which ideas did not. What worked for me may not work for someone else. Yet I have discovered some things in my quest for meaning that I think other people might want to know.

✦

I have had the great privilege of living my life at the intersection of art and science—or, more specifically, at the intersection of evolutionary biology and punk rock. These two fields may not seem to have much in common. When I am teaching biology at UCLA, most of the students do not know that I'm the singer for Bad Religion, though occasionally I see someone with a laptop who is obviously watching one of my shows. And when I am singing on stage, few people know or care about the work I've done in evolutionary biology. But I have found that the two have an underlying connection—a celebration of the creativity inherent in life—that makes the combination less exotic.

Many religious people say that all creation comes originally from God, but I have never believed in God. My parents never pushed the idea on me, and I never saw any reason to adopt it. I have never seen any evidence in anything I have done that any kind of force outside nature acts to influence the physical or biological world. If such evidence could be shown to exist, I would have to consider changing my position. But I do not think any such evidence will be found, and most religious believers do not expect or even necessarily desire such evidence to be found. For me, the existence or nonexistence of God is a nonissue.

If people ask me about my worldview, I say that I am a naturalist. When most people hear that word, they think of someone who spends a lot of time outdoors watching birds and admiring landscapes—and I suppose that description applies to me. But I think of naturalism as a philosophy rather than a lifestyle. From a philosophical perspective, naturalists believe that the physical universe *is* the universe. In other words, there are no *super*natural entities or forces acting on nature, because there is no empirical evidence for anything beyond or outside of nature. Naturalists posit that the universe is made up of only four things: space, time, matter, and energy—and that's it.[6] The matter and energy in the universe can come together in an essentially infinite number of configurations over time, and these configurations cannot be predicted with any certainty for complex systems over extended periods.[7] But matter and energy do not influence and are not influenced by supernatural forces.

I became a punker long before I started thinking of myself as a naturalist, but the two worldviews actually have a lot in common. Punk rock, at its best, embraces an openness to experience, a reliance on reason and evidence, and a questioning of received wisdom. Science, which is based on the naturalist perspective, also is about questioning and not settling for dogma. When a new idea comes along that fits the evidence, we must change our mind about the old ways. If Charles Darwin were alive today, I think he would find something very attractive about punk rock.

Neither punk rock nor naturalism can tell us exactly how to live our lives. They don't answer many of the fundamental questions we confront. Is this person good or bad? How should we behave toward her or him? Where can we turn for guidance? What and whom can we trust? Reasoning plays only part of the role in answering these questions. Much of life is impulsive, instinctive, and seemingly automatic. But understanding what we can control and what we cannot is essential to answering these questions well.

For me, evolution provides the context for our lives. Yes, evolution has implications that can make us deeply uneasy. But on important questions we must seek the truth, even if the truth is difficult to accept. Naturalism can provide the foundation for building a coherent and consistent worldview on which we can base important decisions. In fact, I would contend, it is the only perspective that can secure both our happiness as individuals and our survival as a species.

+

Before the age of fifteen, I didn't understand very much about the world. I lived in one small part of it, in southeastern Wisconsin. My parents were not religious people (though my mother's grandfather Edward M. Zerr was a prominent Bible commentator in the first half of the twentieth century). My mom and dad were both university academics, who steered my brother and me toward Carl Sagan, Andy Warhol, *Monty Python, Saturday Night Live,* and pop radio. I spent most of my childhood wanting to be a singer. I never had more than a B-minus average until late in high school, and even then I was not an accomplished student. Simply put, the "big picture" questions of life never crept into my happy childhood. I was content with the love of my family and my inclusion in circles of good friends.

Mom and Dad separated when I was in the second grade, after which my brother and I adapted relatively easily to living in two households. Mom's new house, in a Milwaukee suburb called Shorewood, became our place for school, homework, and weekday routines. Dad lived in our old house in Racine, twenty-five miles south of Milwaukee. We stayed with him every weekend and most of our summers. We spent hours outside each day on our bikes and playing sports with the kids in the neighborhood—some of whom are still the closest friends I have. My mother and father always got along after the separation, so we were not burdened with the antagonisms of an acrimonious divorce. I may not have recognized my life as idyllic at the time, but in retrospect I realize that it was as close as I can imagine to the great American childhood.

One day, when I was in the sixth grade, my mother sat my brother and me down at the dining-room table. She said that she had gotten a job as an administrator at UCLA and that the three of us were moving to California. My first question was "What about the earthquakes?" She assured us that they didn't happen very often (though I have since been in several). My second question was "What about Dad?" She told us that we would still live with him every summer and during Christmas holidays, just as we had in the past.

When we moved to California in 1976, everything was different. As a junior high school student who had never seen mountains or desert, I found it intensely hot, dry, and foreign. The kids were different from the kids in Wisconsin—way cooler than I was and, at first, much less friendly than my peers in Wisconsin. The girls seemed far more experienced in their sexual sophistication and dress. They paid attention to fashion and ostracized anyone whom they thought was odd.

I had dark brown, wavy hair that was impossible to mold into the popular rock-and-roll hairdos of the 1970s. I wore velour shirts from Kmart and corduroys, because they were less expensive than jeans. I had cheap shoes, also from Kmart or Payless, always worn out, with goofy logos that emulated the popular brands that the other kids wore. I rode a Schwinn ten-speed that was heavy, sluggish, and couldn't jump or skid. I had a powder-blue, plastic skateboard with noisy, open-bearing wheels, totally unfit for the skateboard parks that were so popular in Southern California. I had never in my life been to the ocean. I thought of the beach as a place to go swimming, not as a symbol for a way of life.

When I first got to California, kids asked me, "Dude, do you party?" I thought of our annual kids' New Year's parties back home in Racine. We stayed up past midnight and ate ice cream and drank soda. It took me about six months to realize that "party" was a euphemism for getting high. I saw lots of fellow seventh-graders come to class with red eyes and euphoric smiles and reeking of weed. My classmates in shop class had secretive projects that they brought out only when the teacher took his cigarette break. Their works consisted of salvaged polyurethane cylinders, sealed at the bottom, sanded smooth around the top, with a few quarter-inch holes punched through with the drill press. I was bewildered until a classmate said, "Dude, check out my bong. Bitchin', isn't it." Not only did I not know what a bong was, but I didn't understand the adjective he used to describe it, nor why he was hiding it. All I knew was that there was some weird secret about all this, and I was not privy to the information.

Kids in my school moved up the social ladder by revealing their knowledge of rock-and-roll culture and by sharing their covert collections of black beauties, Quaaludes, and joints. If you partook in their offers, you were one of them, a trusted confidant. If you were afraid to partake, you were a second-class loser. In other words, if you went along with the flow, unquestioning and complacent, you were accepted and rewarded with social status.

If you questioned the norm or went against the grain in any way, you were in for a rocky ride down the social ladder. Despite the mythology that California was free and easy, my school days revealed that there were really only a few acceptable ways to become part of the "in" crowd. I was far more comfortable with my social circles back in Wisconsin, but they had no place at the leading edge of America. In California, I was becoming a loner.

I began to hang out with the people who were labeled geeks, nerds, dorks, wimps, and pussies—or "wussies," the popular way at the time of combining the last two terms. We spent a lot of time listening to and talking about music, a passionate interest I'd brought with me from Wisconsin. But I didn't like the bands favored by the "burnouts," drug users who worshiped Led Zeppelin, Rush, Kiss, Foreigner, Styx, Ted Nugent, Bad Company, Lynyrd Skynyrd, and many others. I had different musical interests.

By the end of junior high, I became drawn toward a new social scene that was just then taking shape in California—the punk rockers, who shunned the mainstream and were despised by most high school kids in L.A. Punk at that time was as much an aesthetic and philosophical stance as a musical genre. It rejected authority of all kinds, which I'm sure is the reason it appealed to me. It also seemed to welcome the anarchy that characterized my life in California. Punk kids seemed to be disillusioned by the promise of California culture. The hopes and dreams of many of their parents didn't pan out, so many were left to fend for themselves as teenagers, because their quiet suburban households were unsupervised most of the time. Divorce was rampant. Many single parents, like my own, had to scramble to earn enough money to pay for the American dream. The kids were expected to behave themselves and find activities that weren't destructive or illegal. The growth of Los Angeles suburbs was so rapid and widespread that the entire region became one great urban sprawl. The great irony of the suburban ideal in Los Angeles is that many of the suburbs have ended up blending seamlessly with the worst aspects of the urban landscape. Teen pregnancy, drug abuse, theft, and the lack of parental supervision—justified by the false premise that suburbs are a safer environment than the "inner city"—led to a dangerous dissonance for many suburban kids. This was the setting that gave rise to the California punk scene. California punk championed a chaotic amalgam of influences, from surf music to reggae, from folk music to pop. It was as diverse as the cultural palette of the entire Los Angeles metropolitan region. The youthful antagonism that arose out of this mash-up seeped into the collective subcultural consciousness of the punk movement. Against the tacit conformity of suburban Southern California, punkers were viewed with both fear and scorn.

I went punk at fifteen. I cut my wavy hair very short, dyed it pitch-black, and stenciled black letters on my T-shirts. I never went so far as to get any piercings or tattoos, but in photographs and videotapes made at that time I do look like a menacing young punk. There were three of us at my high school who were punkers, and all of us, at one time or another, were beaten up by people at school who objected to our looks or our tastes in music. I was mocked daily by a teacher whose classroom was next to my locker. "Punk rock!" he said every day with the same sarcastic, derogatory tone. "Should we all be worried?" That a teacher would pick on me was a completely foreign concept. My parents were educators who never would denigrate someone for expressing individuality.

The violence from other students scared me and at the same time made me feel powerful. It made me realize how frail most of the conformists really were, how easily they could be pushed to the point where they could lose control. I found great solace in the community of punkers from different schools, different neighborhoods, and different cultures, all with similar stories of oppression and abuse. I began to feel that there was a way to deal with the social alienation of my West Coast surroundings. It was through questioning and challenging, not conforming and accepting. Learning to be an individual was the best gift I got from growing up punk.

By the time I entered tenth grade, I was completely alienated from the scene at school. I lived only for the summers I could spend with my friends back in Wisconsin. In L.A., I passed most of the time by myself or with the few fellow outcasts who endured the daily barrage of punk insults.

Two things in high school saved me from drifting into a meaningless existence. The first was music. I made friends with a kid two grades older than I, named Brett Gurewitz. He was a punker, like I was, not interested in school but fiercely smart. He wanted to start a band and had guitars, microphones, and a PA system but no confidence in his singing. A mutual friend introduced us by saying, "Hey, Brett, this is Greg—he's a great singer." Actually, at that point, I had never held a microphone in my life. But I had somehow convinced my friend that I was an experienced singer, and within the week Brett and I were writing and playing songs together as guitarist and singer. Brett knew a drummer, Jay Ziskrout; the first time we rehearsed together, I sang at the top of my lungs, and we all agreed that we were so good we would meet again in a week. We talked another friend of ours,

Jay Bentley, into switching from guitar to bass. Within a month, we were a four-piece ensemble with six songs that Brett and I had written. We began rehearsing every day after school in my mom's garage—a sweltering, dark, detached building away from the house that became affectionately known as "the hellhole."

We then faced a common problem for new bands—what to call ourselves. Many people ask us about the name of our band, which we eventually chose after many brainstorming sessions. First of all, you have to remember that we were fifteen-year-old punks—we wanted to piss people off. Anything that might make parents, teachers, and people with authority bristle was up for discussion. We also wanted a name that would suggest a great logo for stickers and T-shirts. Many of the names were compelling but too repulsive. Smegma, Vaginal Discharge, and Head Cheese might make for great logos but were quickly rejected as not representative of our songs. We played around with a lot of names involving the word "bad"—Bad Family Planning, Bad Politics. When we hit on Bad Religion, it seemed perfect. That year, 1980, was a time of rising prominence for televangelists like Jimmy Swaggart, Pat Robertson, and Jim Bakker.[8] The year before, Jerry Falwell had founded the Moral Majority, which was having a powerful influence on the presidential election between Jimmy Carter and Ronald Reagan. Religion was a hot topic, and those TV preachers seemed like a good target to us, though we didn't think they could possibly last for more than a few years. We knew that most people were so defensive about their religious ideas that they would be highly offended by our name—a major plus! And then Brett came up with a logo that represented our philosophical stance. We felt complete.

Thirty years later, the name Bad Religion continues to offend some people. So does our logo—a Christian cross with a red slash through it (which has become known as the Crossbuster). But none of us has ever regretted choosing that name or the logo. We consider the Crossbuster a kind of "no parking sign," in the sense that if you fly the logo it means "you won't find Christianity here." And the name and logo established us from the beginning as a band that was willing to think for itself. It suggested that our songs would have a more philosophical edge. It kept us from having to deal with people who were unreasonably critical of music that challenges convention. In short, it gave us a creative freedom to do what we wanted.

The other thing that saved me in high school was my discovery of science and especially evolution. For Christmas in 1977 my mom gave me a book called *Atlas of Early Man* by Jacquetta Hawkes, which I still have on my bookshelf.[9] It traces the history of human beings from 35,000 years ago to the present, with lots of pictures and timelines to illustrate important developments. I read that book carefully in junior high. I couldn't understand a lot of it, because it included concepts that were too advanced for me at the time. But the overall narrative was so compelling that I was able to piece together a basic chronology of human cultural history.

Two Christmases later, my mom gave me the book *Origins,* by Richard Leakey and Roger Lewin, which describes the evolution of human beings from our apelike ancestors.[10] That book introduced me to a much richer history of our species. This was about the time when Brett and I were forming Bad Religion, and the lyrics of one of the early songs I wrote as a sophomore in high school were inspired by that book. Here are the last lines of *Origins:*

We are One People, and we can all strive for one aim: the peaceful and equitable survival of humanity. To have arrived on this earth as the product of a biological accident, only to depart it through arrogance, would be the ultimate irony.

The song I wrote is called "We're Only Going to Die from Our Own Arrogance," which we included on our first record in 1981, just after I turned sixteen. Here are the lyrics:

> Early man walked away as modern man took control.
> The minds weren't all the same, to conquer was his goal.
> So he built his great empire and he slaughtered his own kind.
> Then he died a confused man, killed himself with his own
>   mind.
> We're only gonna die from our own arrogance.

That song became a defining statement for Bad Religion. I still sing it at concerts today, and other bands later paid tribute to us by recording it on their own albums. Looking back, what that song demonstrates to me is that I was already committed to seeing the world through the lens of evolution. I was far from developing a coherent naturalistic worldview. But reading about science, and writing about it in music, proved to me that there is great compatibility in the arts and sciences. I wanted to be taken seriously as a songwriter, and the scientific worldview could be my inspiration.

I had no fashion sense. I did not have unique looks or cool hair. But I had started to develop a unique worldview from studying evolution and biology. I put all of my effort into developing the idea that if I could sing about unique subjects, then it really didn't matter if I wasn't cool or whether I took drugs. I could carve out a niche for myself as something unique in the punk world: a singer of lyrically rich songs with an undercurrent of natural science.

✦

One place to learn about evolution was in school, but I didn't get much help there. As is the case with many high school biology classes, my school downplayed evolution; though it is the key to all of biology, we got only a one-week unit on the topic. So I had to educate myself. I bought a cheap paperback version of *On the Origin of Species* and set a goal of reading some of it each night before bed. I began putting together a library on evolution that today occupies an entire room of my house.

For our final project, we were supposed to come up with some sort of presentation about what we had learned that year. Instead of regurgitating one of our lab experiments, as most of the students did, I wanted to mimic Donald Johanson, the famous paleoanthropologist. He had given an inspiring public lecture that I attended, where he talked about the discoveries he had made in Africa, including his most famous fossil find, Lucy.[11] I borrowed the El Camino Real High School slide projector, took photographs of color plates from my books, and put together what must have been the most rudimentary explanation of human evolution ever presented. I explained to my classmates that evolution was based on competition and that some forms of life were better at living than others. I told the class that all evolution tends toward perfection, and that, despite numerous false starts and dead ends, the most successful and elaborate evolutionary lineage was the human species. I said that all human attributes were originally adaptations to life on the savannah in Africa. In short, I described a just-so story of human existence, heavy with the tacit suggestion that life's purpose was simply for all species to evolve into more highly advanced and perfect individuals.[12]

Much of what I said in that lecture was wrong. Evolution does not tend toward perfection. It depends as much—if not more—on cooperation and random chance as on competition. Evolution does not have a direction. It is anarchic, yet out of this anarchy have come biological entities of great sophistication and beauty. Many of our most important human features are not adaptations to prehistoric environments, and humans are far from evolution's crowning achievement.[13] I'm sure most of my classmates were happy to take a nap during my talk. But I received an A in that class, and my teacher wrote on my report card "Gave a great talk on evolution."

I'm still proud of that grade, but I remember thinking at the time that there was a lot I had to learn. Forming Bad Religion and discovering evolution forced me to consider the important issues in life. But to sing about the issues of daily life from a perspective that included science, I knew that I needed a much larger vision of the world.

*

One of the most famous statements ever made about evolution is "nothing in biology makes sense except in the light of evolution."[14] In high school, I interpreted this statement to mean that we can't understand life if we don't understand evolution. I was sure that evolutionary science could answer the great intellectual questions of my adolescence.

I still believe that everyone needs to have at least a cursory understanding of evolution. Some people are deeply disturbed by the implications of evolution, but it is an essential part of the narrative of the modern life sciences. I cannot agree with those who reject evolution in favor of a creationist philosophy, especially if they base their rejection on the grounds that religious authority should take precedence over science. People need to understand the basics of evolution if they are going to reject it—otherwise, they are not contributing anything productive to modern society. To me, a refusal to accept evolution amounts to denying all the progress made by twentieth-century science and returning to a time when creationist natural theology, consistent with church authority, was the best source of information about the natural world.

The mechanisms involved in evolution are different from the events that give rise to my feelings. Still, reading about evolution has helped me through many difficult times. Evolution is full of dead ends, as is the experience of every human life—relationships ended, time wasted, songs unfinished, goals not achieved. All organisms and species die, just as all of us must die. Yet tragedy also entails creativity and opportunity. Over the course of deep evolutionary time, species go extinct and the earth is repopulated with new ones.[15] As Michael Crichton's chief scientist in *Jurassic Park* says, "Life finds a way."

I have never given in to nihilism, as did some of the other punk rockers I knew. I never subscribed to the philosophy that "everything sucks, so why bother." People who think that self-destruction is the only logical response to a world without God are not noble realists—they are misguided at best or, more severely, mentally ill. What we do has a profound influence not just on those closest to us but on much wider circles of people than we imagine. Having a role in the narrative of life on this planet gives me a sense of place. It gives me a perspective from which to view bad situations. It helps me recognize my importance to those around me and the need to enhance rather than diminish their lives.

Can naturalism compete with religion in providing a basis for a meaningful life? I believe it can. Naturalism is not a religion. It does not presuppose a world beyond the world that we can witness empirically, as most religions do. But naturalism can offer a template for a meaningful, internally consistent worldview. At the very least, an understanding of evolution can offer a basis for coming together as rational beings to agree on the answers to difficult questions.

CHAPTER 2

MAKING SENSE OF LIFE

Do I think it is useful to study evolution? I think the answer
is yes, because the worldview that we want, what I want, is
a materialist worldview. Everything true that we can learn
about nature adds to our understanding of the material world,
and that's desirable.

— Richard C. Lewontin[1]

Even though two viewpoints seem to be completely
incompatible, some people in their thinking and their feeling
can bring the two things together and believe simultaneously
in both of these things. Now, don't ask me to explain that. I
consider this something that cannot be explained.

— Ernst Mayr[2]

Darwin's legacy is that of the connectedness of life through
time.

— Lynn Margulis and Dorion Sagan[3]

The first time I sang punk rock at a show, I found myself nervously daydreaming of Wisconsin, two thousand miles away, where my best friends were going about their daily lives oblivious to the social suicide I was about to commit. I was on a makeshift party stage in a Santa Ana, California, warehouse.

Brett and Jay were holding their instruments on either side of me. Our original drummer, Jay Ziskrout, was sitting forlornly behind us on the stool of his drum kit. Someone's dad owned a small company that shipped and stored canned goods, and the dad had given his daughter permission to use the warehouse for a small birthday party for her and her punk-rock friends. More than two hundred celebrants from all over Los Angeles showed up with their mohawks, boots, bandannas, chains, and bad attitudes to discover that the rumor of free beer was a sham. Now I had to sing, the first time ever in public, for a very angry crowd.

"We are Baaaad Religion." Those were the first words I ever uttered at a show. I was bent over and looking straight at the tips of my Vietnam-era, army-issue combat boots. I have never minded being the front man for a punk-rock band. I enjoy provoking people with my songs. But it's one thing dreaming about it in your garage and another thing being up there on stage. I was scared shitless.

I paced back and forth from side to side like a speed skater, never looking anyone directly in the face. No one knew where the birthday girl was. In fact, no one in the band knew her, and not one of us knew how we got invited to this gathering. All I knew was that we needed to play our eight songs as loudly and as quickly as possible and try to get out of there without getting our asses kicked. After all, we were from the San Fernando Valley, a place with a reputation for high school football and weekend garage sales rather than punk bands. The only bands that actually were from "the Valley" tried desperately to keep their geographic origins secret.

When we began the first song, even the punkers were startled by the feedback and oversaturated distortion from Brett's guitar.

I remember feeling some relief as the attention shifted away from me for a moment. Maybe Brett could absorb their scorn. Then it was time for me to begin singing. As the first words came out of my mouth, I had the surreal feeling that time had stopped. I had worked hard on the ideas and concepts in our songs, but none of that mattered anymore. Now I had to perform.

My voice could barely be heard through the dissonance created by the overdriven amplifiers and toylike PA speakers, and I had no idea how the crowd was reacting. But after a few lines I finally had the courage to raise my head and glance out—and I beheld a miracle. The anger of the audience had turned into an ecstasy of flailing limbs and jerking heads. They were slamming senselessly into one another, wild-eyed, shouting along to the beat. The music we made had generated a disjointed, aggressive, emotional unfolding of collective motion. I immediately felt a strange sense of security. Whenever I sang, the crowd would react. Against all odds, as a fifteen-year-old high school sophomore, I had become the center of this delirious, whirling, cacophonous chaos.

But I would be condemned as a charlatan if I showed pleasure in any way. So I spoke very little between our songs, and by our third song I had it down. Middle-finger attitude and speedskater posture—I had discovered a formula that would work for years.

By the end of our performance I had enough confidence to say, "Next up, Social Distortion." But the reaction of the crowd was muted, and for a moment I thought I had mispronounced something. But it turned out that the punks just wanted to hear more fast drums and distorted guitars—they didn't really care who was playing it. Still, I was as sure as a fifteen-year-old can be that the audience wanted more of what Bad Religion could deliver: multisyllabic, difficult-to-understand lyrics about God, evolution, and life's big questions. And I had enough belief in my delusion to make Bad Religion the focal point of my life.

✦

We formed Bad Religion during a pivotal time in the history of American rock and roll. Old musical forms were faltering, which opened up tremendous opportunities for new kinds of music. "Classic" rock bands like Aerosmith, Journey, and Kiss were filling arenas, but their music was brain-dead—almost a parody of itself. Disco had come and gone without leaving behind a single notable band. Progressive rock, after some promising experimentation early in the 1970s, was collapsing under the weight of its own ponderousness.

The rise of punk music in the mid-1970s was a result of several forces.[4] It was partly a reaction to the bombast of mainstream music, partly a return to rock-and-roll fundamentals, and partly a musical and philosophical statement of independence. Punk was centered in three hotbeds of activity: England, New York, and California. By the time we formed Bad Religion, bands like the Ramones, the Dead Boys, and Blondie were playing at CBGB's, Max's Kansas City, and other clubs in New York. In England, the Sex Pistols had self-destructed, while Sham 69, the Clash, and the Buzzcocks had become famous enough to headline their own tours in the United States, where they had a big influence on younger bands. In California, the punk scene was more diverse, with highly influential bands like the Weirdos, Black Flag, the Circle Jerks, X, the Germs, the Gears, the Dickies, and Fear. Before 1981, the L.A. punk scene was an interesting blend of styles.

In the movement's infancy, a punk club could feature art-rock bands like Geza X and the Mommymen opening up for a rocka-billy band like the Gears, with a hard-core punk band like Black Flag headlining, all in the same evening. It was a tolerant and welcoming scene in those days, with people finding many ways of expressing their punk lifestyles. It wouldn't stay that way.

Brett and I grew up listening to many different types of pop music, from "prog-rock" to Top 40, and all of those styles in-fluenced our songwriting. But we thought of ourselves as punk rockers and we wrote punk-rock songs. We were especially influ-enced by the harder-edged punk bands that paid careful attention to pop-song structure, like the Dickies, Buzzcocks, X, Sham 69, the Ramones, and our contemporaries, the Adolescents. The use of poetic lyrics by singers like Elvis Costello and groups like the Germs inspired us to focus on the meaning in our songs. We even borrowed formulas from the Beatles, Elton John, and Todd Rundgren, though we never would have admitted it at the time, because of the scorn punks heaped on non-punk music.

The nonconformist slant of our songwriting reflected, in part, events during that period in American history. The nation was becoming politically more conservative and conformist. Southern California was a hotbed of televangelism and right-wing, small-town politics. Punk music gave us a way to rail against the dead-ening groupthink of the suburbs while offering an alternative to the political mainstream. In that respect, the punk movement in Southern California grew out of suburban discontent, as opposed to the English punk scene, which reflected the disgruntlement of the working class, or the New York scene, which was a purely ur-ban counterculture rebelling against prevailing artistic standards. The focus on the suburban lifestyle in Southern California punk set it apart from the punk in other cities, and that's partly what made it so influential in shaping later rock music.

Even though our music was often angry and despairing, we felt a sense of great opportunity. In 1980, we found ourselves in the midst of a new, developing music scene that stood in willing contradiction to the tired hippie ideals of the previous decades. It was a time of broad social change, and we wanted to join in the fray. In retrospect, I can't say that we recognized the significance of the changes that were taking place, for better and worse, in society at large. But the punk scene, to us, was life-sustaining. We had found a way of expressing ourselves that connected us with other people caught between the dying youth culture of the 1960s and 1970s and the uncharted, ominous decades leading to the new millennium. We believed music would provide the social cohesion needed to create a new vision of the future.

This brief overview of Bad Religion's origins can't help but sound familiar to me as someone who has studied evolutionary biology. Evolution happens when populations of organisms take advantage of what, seen in retrospect, are tremendous and usually unexpected opportunities. Populations of organisms have no conception of opportunity, as we did when we started our band. But in the course of natural history, seemingly small innovations can sometimes have global repercussions. For instance, more than a billion years ago, one single-celled organism began living inside another single-celled organism.[5] This mutualistic relationship was so successful that today the cells in every multicellular organism are descended from this innovative ancestor.[6] Sometime before 375 million years ago, a species of fish began to spend more time on land, possibly in search of prey or to escape the dangerous predators of Paleozoic seas.[7] That fishlike vertebrate was the ancestor of every four-limbed land animal that has ever existed, including us. Beginning around 100,000 years ago, a small group of gracefully built, exceptionally brainy humans began to expand outward from their homeland in eastern Africa into territories occupied by other groups of humans, including the Neanderthals in Europe and a species of humans named *Homo erectus* in Asia.[8] Today all humans are descended from that small group of eastern Africans while the Neanderthals and *Homo erectus* died out long ago. At each transition, the ancestral populations anticipated nothing about the consequences of their actions. They were simply reacting to the environmental conditions of the time. Yet they set in motion profound changes that remade the world.

I don't want to take the parallels between biological evolution and cultural evolution too far. They are very different processes and have very different results. Some evolutionary biologists forcefully resist interpreting human affairs in evolutionary terms. When I was working on my PhD, I interviewed twelve prominent evolutionary biologists in the United States and England, and one of them, George Williams of the State University of New York at Stony Brook, falls squarely into this camp. "If it's natural behavior, it's bad, it's evil," he told me.[9] My PhD adviser, William Provine of Cornell University, also dislikes drawing comparisons between evolution and cultural change. "Evolution is not my friend," he wrote to me. "Evolution cares nothing about me. Meaning in my life comes from people who care about me."[10]

I can't go so far as to characterize evolution as evil. Evolution is simply the way the biological world works, whether we like it or not. And once the reality of evolution is accepted, it has a strange and forbidding beauty. It occurs over time periods beyond human comprehension. It has created organisms of fantastic complexity and immense order. Of course we must resist committing what philosophers call the "naturalistic fallacy"—the idea that we can draw ethical conclusions from the workings of nature.[11] Biological evolution does not justify cruelty to other humans in a misguided effort to get ahead. Nor does it justify oppressive social institutions based on parodies of evolutionary ideas. Still, properly understood, biological evolution can be a rich source of insight into questions we face every day. And for me, as a kid growing up in California, biological evolution satisfied my curiosity and provoked me to look deeper into life's big questions. It was far more satisfactory for me than anything I heard from political, religious, or cultural leaders. Whenever I thought about my own life, I inevitably was drawn toward evolutionary analogies.

Cultures transmit traditions, ideas, words, and music from one generation to the next, and cultures gradually evolve in the process. But there is a big difference between cultural evolution and the biological evolution of organisms. For one thing, organisms do not evolve biologically over the course of their lifetime. They change profoundly, as when a fertilized egg cell grows into an adult human. But that is not biological evolution; it is biological development. Evolution is a process that occurs in populations of organisms over multiple generations. A population evolves when individuals with particular traits die off and are replaced by offspring with different traits. That's one of the things that makes biological evolution both momentous and fearsome. It is literally a matter of life and death.

If evolution consists of a change in traits over multiple generations, then what's a "trait"? It can be many things. A trait can be an anatomical feature, such as the size or shape of a limb, the coloring of skin or fur, or the number of petals on a flower. A trait can be a behavior, which, in turn, may reflect the anatomical connections of an animal's brain cells. Or a trait can be purely biochemical, such as the compounds circulating in blood or the molecular composition of a skeleton.

Traits have many origins. Today people tend to think of traits as arising from "genes," which are encoded in the DNA molecules in our cells.[12] However, genes are not the only source of our traits. The egg and sperm cells from which we all developed contained many molecules besides DNA, and these molecules exerted an influence on our developing bodies. Also, our bodies were influenced by the nutrients, toxins, and even sounds that we were exposed to in the womb. Most important, as soon as we were born, the biological molecules in our bodies began to engage in constant interactions with an immense variety of factors in the environment, from the composition of the air we breathe to the conversations we have with others. These days, when someone asks me if I favor "nature" or "nurture" in the timeless debate about which is more important in determining human traits, I give the only answer that makes sense: I am an interactionist.[13] The traits of individuals are the result of ongoing interactions between their biological molecules and the environments they encounter throughout their lifetimes.

Biological evolution requires two things. First, offspring have to differ from their parents. Such variation arises naturally in sexually reproducing organisms like ourselves. The mixing of biological molecules, including DNA, from two separate organisms generates unique combinations of traits. Also, these biological molecules can change from one generation to the next, as can the environment, and these factors interact to produce differences in offspring. Think about how different you are from your parents—or how different you are from your siblings, although you have the same biological parents. Even in species that reproduce asexually—such as single-celled organisms that simply divide in two—traits gradually change through alterations in biological molecules and in the interactions those molecules have with the environment.

The second requirement for evolution is that a trait be biologically heritable. For example, traits that are encoded in DNA can be passed down to descendants through egg or sperm cells or when a cell divides. But genetic inheritance is not guaranteed. For instance, a trait may be encoded on a piece of DNA that fails to get distributed into a sperm or egg cell. Or the environment may change in some way to prevent the trait's expression. The passage of traits in DNA from one generation to the next is an essential component of evolution, but it is not the only way traits are transmitted.

Evolutionary biology historically has focused on a particular kind of change in heritable traits. Some traits enable an organism to have more offspring than other organisms in a population. In this way, a heritable trait can become more abundant in the next generation. It's a simple numbers game. New traits appear first in a single organism (like the debut songs on a pioneering punk album). But they can appear in increasing numbers of organisms with each new generation if they help organisms have more offspring than others in the same population (as those songs inspired other songwriters to form bands of their own). After enough generations, a trait can become so widespread that it is essentially universal within a population of organisms (as when punk became mainstream and punk songs were heard, as they are today, on commercial radio). By the same token, if a new trait causes an organism to have fewer offspring, that trait is unlikely to persist (like so many failed experiments in punk music, such as Bad Religion's "lost album" *Into the Unknown*).

In the previous paragraph, I was drawing analogies between biological evolution and the history of punk music. But again it's important to note that the two processes are quite different. The most widely accepted view of biological evolution is that the gradual accumulation of traits is due to some organisms in a population leaving more viable offspring than other organisms. In this way, populations of organisms gradually become more adapted to the environments in which they live. The punk scene evolved not from heritable variation but rather from cultural innovations that struck a nerve with willing groups of misfits. Still, it's hard for me not to draw evolutionary parallels. I used to envision each Bad Religion concert as a unique environmental opportunity. We could try to increase our popularity trait by singing better songs and giving better performances, in which case our popularity would grow. Or we could suck and lose fans, causing eventual extinction. Either way, the similarities seemed obvious to me.

My resistance to authority eventually carried over into my science. In graduate school, I once did some research related to the evolution of fishes. The general consensus among evolutionary biologists is that fishes originated in salt water, probably in the shallows near the shore. Many renowned scientists support this consensus, but hardly any of them have done any geological work on the sedimentary rocks in which the earliest fossils of fishes are preserved. My graduate adviser,[14] recognizing my antiauthoritarian youthfulness, knew that this was the perfect project for me. I could produce some basic data that would cause a stir among the gods of the paleontological community.

A small contingent of scientists offered an alternative hypothesis that fishes originated in freshwater lakes and streams. But they staked their claim on comparative anatomical and physiological data, not field-derived geological work. To test this hypothesis, fossilized fish needed to be carefully studied. If the rocks surrounding fossils of very early fish could be analyzed, perhaps the environment in which those fish lived could be determined.

For a couple of summers I worked in the Sangre de Cristo mountains of Colorado. High above the tree line are sedimentary rocks that contain some of the world's oldest fish fossils. We were twenty miles away from any sign of other humans, high above the Colorado forests, with a panoramic view of the San Luis Valley almost nine thousand feet below. We camped for weeks at a time with all our equipment carried in on horseback. At the beginning of the summer, my field adviser, Ted, pointed to one side of the immense valley where we were standing. "That's your area to map. I'll take the other side. Meet back at camp for dinner."

I collected sediment samples and tiny bone fragments all day long. For lunch I ate granola bars and beef jerky on windswept rocky promontories. I would go ten hours at a time without saying a single word to anyone. It was a lonely pursuit, but giving my voice a rest was pleasurable. We singers have to use our voices in unnatural ways, and sometimes not saying anything at all can be a great relief.

Despite the loneliness, I was excited by the science. I was collecting fragments of the earliest vertebrate hard tissues, the first organisms with a bony skeleton. I put the fragments in small canvas sacks. Out west, at that time, you could go into small-town banks and tell them you were collecting rocks and fossils, and they would gladly sell you any surplus money sacks they had. I still have samples from my fieldwork in those bank sacks.

When I got back to the lab at UCLA, I analyzed the sedimentary rocks under my microscopes. Much to my satisfaction, the research did not support the reigning consensus. On the contrary, the oldest fossils of fishes appeared to be embedded in rocks that originated in freshwater, because the rocks were devoid of any marine invertebrates characteristic of saltwater environments. Perhaps more important, the sediments that hold these early vertebrate fossils contain characteristic marks, or "signatures," associated with river systems, not with shallow, offshore environments.

My study was not widely recognized, despite its publication in a major paleontological journal.[15] But the field of paleoichthyology (fossil fish studies) has continued to progress. Today, there are older fish fossils than the ones I worked on, but the question of whether vertebrates originated in the sea or in streams remains open. As in all branches of science, discovery leads to a more detailed and satisfying story of the past while simultaneously creating new questions that call out for answers. And I have never forgotten that engaging in scientific data collection can be a great way of resisting authority.

Given my history in this field, it has been very interesting for me to read about a seminal field study that has been conducted over the past few years by a team of paleontologists from the University of Chicago, the Philadelphia Academy of Natural Sciences, and Harvard University.[16] The researchers were looking for fossils on Ellesmere Island, high above the Arctic Circle in northern Canada. They could only work for a few weeks each summer after the previous winter's snow melted and before the first fall storms, and they often had to endure high winds and freezing temperatures. But for a fossil hunter like me, the landscape is perfect. The ground is entirely bare of vegetation, and the rocks are broken up by the continually freezing and thawing temperatures. Fossils can weather out of the surrounding stone like a swimmer emerging from under water.

They were looking at rocks formed from sediments that were deposited about 375 million years ago by meandering streams that flowed into the ocean. At that point, the landscape looked very different than it does today. Because of continental drift, what is now Ellesmere Island was then located near the equator. The ground was covered by primitive plants, including ancient ferns, horsetails, and the oldest seed-bearing plants. Ancient sharks and fishes filled the oceans, but the only animals that had been living on land for extended periods were arthropods like mites, scorpions, and other very early insects.

At the very end of the 2004 field season, the scientists found what they were looking for. It was the fossil of a fish with scales, fins, a flat head, and eyes on top of its head. But it had a neck, so it could move its head independently of its body, which is something that the fish living today cannot do. It also had lungs, as do some fish today, so it could breathe air directly from the atmosphere rather than relying on shallow, oxygen-poor water. Most important, its front fins had one large bone connected to two smaller bones connected to a jumble of bones at the end, which is the same arrangement as the bones in our own arms. Biologists call this arrangement "epipodial" and "propodial." Your epipodialia connect to your feet and to your hands; in your legs, your epipodialia are called the tibia and fibula, and in your forearms they are called the radius and ulna. Your propodialia connect your epipodialia to the rest of your body. In humans the propodialia are the femurs of your legs and the humeruses of your arms.

The fossil found on this Arctic expedition was fishlike in almost every way. But its limb structure was very suggestive of the earliest-stage terrestrial vertebrates and was not fishlike at all. The scientists named the fossil *Tiktaalik,* which means "freshwater fish" in the language of the native Inuit.

Tiktaalik is a perfect example of a transitional fossil between two evolutionary lineages. It has many characteristics of the fish that then occupied the oceans. But it also has characteristics of the amphibians that within a few million years would be spending much of their lives on land. Amphibians would eventually give rise to reptiles, which would give rise to mammals, which would give rise to primates, which would give rise to an unusual primate known as *Homo sapiens.* That's right. Humans are descendants of this very unlikely sequence of vertebrate lineages, each one branching off from an ancestral stock.

How can we describe *Tiktaalik*'s evolution from previous species? Evolutionary reconstructions, like all historical accounts, almost always have an element of speculation, but a plausible account goes like this. Sometime before 375 million years ago, a species of fish must have lived in the vegetation-choked, freshwater streams of a tropical landscape. As part of the natural variation of traits among individuals, some of the members of this ancestral species must have had longer, more articulated front fins than other members of the species. These fins may have given those individuals some kind of advantage in the streams where they lived. *Tiktaalik* had powerful muscles in its front fins, suggesting that it used its fins to push itself above the water. Perhaps an individual with longer and stronger limbs was better able to look for prey, so that it was a better hunter than others in its cohort. (*Tiktaalik* also had rows of sharp, pointed teeth, indicating that it was not a vegetarian.) If so, individuals with stronger limbs could have had more offspring, and many of those offspring would have had the more developed limbs of their parents. Meanwhile, some of these offspring might have had other traits well suited to hunting prey on land, such as better vision and greater mobility, and a circulatory system more suited to living on land. Over thousands of generations, these traits would accumulate in the descendant populations. New variations might be produced as pioneering populations spread out into various microhabitats in the untrammeled terrestrial environment. Eventually, groups with these traits would be quite different from their ancestors that lived in the oceans. With reproductive isolation, separate and distinct new species would form.

Speciation is especially momentous when a species stumbles upon a new way of life, as *Tiktaalik* did when it spent more time above water and less in it. Suddenly, untapped resources were available in a landscape devoid of predators. After the successful first excursion to new ways of life, further speciation can occur very rapidly, as new species originate to take advantage of new opportunities. Within a few million years of *Tiktaalik,* many species of amphibians were spending much of their lives out of water. Eventually, the separate evolutionary lineages of amphibians began to resemble one another less and less. Distinct traits beneficial to unique ways of living on land accumulated, leading to more new species that differed from the ancestral species.

This kind of pattern, where one ancestral species gives rise to a plethora of new species, is called an "adaptive radiation."[17] New traits evolve in response to the colonization of new habitats. The greater the variety of habitats that a species encounters as it spreads into new geographic regions, the more likely that it will split into a large variety of descendant species over many thousands of generations. And the move to land was especially fruitful. From the first fishlike animal that made the transition to land evolved everything from dinosaurs to reindeer, elephants to hummingbirds, snakes to human beings.

The origin of new species from preexisting species leads to the familiar "tree of life" diagram.[18] Every species alive today is represented by the tips of the smallest branches at the top of the tree. But each species is descended from an earlier species, leading back in time from the twigs to the branches to the trunk of the tree. Thus, every species is related to every other species. Some species are more closely related, as with humans and chimpanzees. Some are more distantly related, as with humans and slime mold. But every pair of species living today shares similarities, because they are both descended from an ancestral species, although you might have to travel far into the past to find it.

It was just a little more than a century and a half ago when Charles Darwin and Alfred Russel Wallace explained how all the species on earth could be the product of ancestor-descendant relationships.[19] It was the most important discovery that has ever been made in science—more important than the discovery that Earth revolves around the sun, more important than the realization that the universe is billions of years old, more important even than the discovery of the constituents of the atom. The recognition of biological evolution completely overturned how humans thought about themselves and their relation to the rest of the universe. It's no wonder that many people still have not reconciled themselves with the implications of evolution.

To recast the quotation I cited in chapter 1, before Darwin and Wallace published their ideas in 1859, nothing in biology made sense except in the light of natural theology. Prior to Darwin's and Wallace's announcement, most of the people whom we would today describe as scientists studied nature as a way of revealing the divine intentions of a purposeful god. Some were called "natural philosophers," and others were called "natural theologians." They interpreted the distribution and characteristics of living things as evidence of God's willful design of creation. Species, to the natural theologian, were "thoughts in the mind of God," created in their current form and current location for reasons known only to God. In 1829, as he was lying on his deathbed, the Reverend Francis Henry Egerton, the eighth and last earl of Bridgewater, commissioned a series of books on "the Power, Wisdom, and Goodness of God, as manifested in the Creation." The books were supposed to use "the variety and formation of God's crea-tures," "the effect of digestion," "the construction of the hand," and "an infinite variety of other arguments" to illuminate God's work in the design of nature. Eight of these so-called Bridgewater Treatises were published between 1833 and 1840.[20] For example, in his 1834 book *The Hand: Its Mechanism and Vital Endow-ments as Evincing Design,* Sir Charles Bell began by writing: "If we select any object from the whole extent of animated nature, and contemplate it fully and in all its bearings, we shall certainly come to this conclusion: that there is Design in the mechanical construction, Benevolence in the endowments of the living properties, and that Good on the whole is the result."

Darwin's and Wallace's evolutionary theory turned natural theology on its head. Their explanation of the biological differences between species showed how natural processes could produce all of the species on earth without the need for divine intervention. The hand, for instance, is not a device constructed by God to meet the needs of humans. It is rather a modification, over millions of generations, of the fin *Tiktaalik* (or *Tiktaalik*'s close relative) used to push itself out of the water.[21] The role of God in the pageant of nature was reduced to irrelevance after the publication of Darwin's *Origin of Species*.

The discovery of evolution had a profound effect on the way people view the world—though its acceptance has been a slow, multigenerational process. In the middle of the nineteenth century, it can safely be said that most biologists were religious believers. Darwin himself prepared for the ministry at Cambridge before his five-year round-the-world trip on the *Beagle*, which led him to begin questioning the providence of nature's design. Darwin delayed publishing his findings for more than twenty years, because he knew they would be socially explosive (and because he didn't want to offend his devout wife). Even after Darwin made his evolutionary ideas public, there was considerable reluctance to explore their implications fully. As the Scottish writer Thomas Carlyle once said to the English biologist Thomas Huxley, "If my progenitor was an ape, I will thank you, Mr. Huxley, to be polite enough not to mention it."[22]

In the early part of the twentieth century, the religious beliefs of scientists themselves became an object of study. In 1914, James Leuba, an American psychologist at Bryn Mawr College in Pennsylvania, polled four hundred scientists identified as especially prominent in the 1910 edition of *American Men of Science*.[23] Leuba asked them whether they believed in a god "to whom one may pray in the expectation of receiving an answer." He also asked them whether they believed in immortality, or life after death. He found that about a third (32 percent) believed in a personal god. Slightly more (37 percent) believed in immortality.

In 1933, Leuba repeated basically the same questionnaire. Belief in both a personal god and immortality had dropped dramatically—to about one in seven.[24] Leuba predicted that belief in a personal god and immortality would continue to drop among the most respected members of the scientific community.

I became aware of this project as a PhD student when my adviser, Will Provine, said, "You've been writing music and singing about bad religion for years now, and you have a strong background in evolution. Why not do a project on religious belief in evolutionary biologists?" I immediately saw what a great suggestion this was. It would allow me to explore the links between religion and evolution while discussing with authorities what they thought about evolution's impact on traditional religious belief. I polled 271 professional evolutionary scientists elected to membership in 28 honorific national academies around the world. Of them, 149 evolutionists from 27 countries answered my questionnaire.[25] I also conducted personal interviews with 12 of them to gain deeper insights into their thinking.

One of the highlights of my project was an afternoon talk with Richard Dawkins in Oxford, England. We sipped tea in his "garden," which in England means "out on the back patio." He said that he admired aspects of my questionnaire and thought it a worthy study, which was a great affirmation for me. No matter how widely the results are cited in the field, I consider those data as a benchmark measuring the degree of compatibility between evolution and religion among the last generation of great twentieth-century evolutionary biologists.

My overall findings confirmed Leuba's prediction. Only 13 of the 149 respondents—about 9 percent—believed in a god who plays an active role in the world. Other polls conducted in recent years have shown comparable results.[26] Belief in a god who exerts an influence on the natural world has not dropped to zero among prominent biologists, but believers in this group of scientists have become a small minority.

In my poll, I also wanted to distinguish between two kinds of religious beliefs. One group of religious people believes in a god who created all of the matter and forces in the universe but does not intervene in daily events. After the moment of creation, God stepped away and let his creation run on its own. This belief in general is known as deism, though there is considerable variation among deists. Some simply believe that God created the universe and then stood back to watch what would happen. Some believe that God organized the world in such a way as to produce humans. Others see in human morality an argument for God's existence. Some deists may even believe in life after death. But deists generally reject the idea of a god who answers prayers or intervenes in human affairs.

More traditional believers worship a god who intervenes in their daily affairs. This position is known as theism. The great majority of religious believers are theists. They believe in a god who answers prayers, cares deeply for the well-being of all, and causes things to happen for the benefit of someone or something. Theists have a personal relationship with their god, and most of them believe adamantly in an afterlife. Not surprisingly, they tend to view deism as insulting, because deism denies the existence of a personal, caring god.

In my poll, I allowed respondents to identify themselves as adhering to a single belief or a combination of beliefs. Only 2 of the 149 respondents described themselves as full theists. None identified with pure deism. And 116—almost 80 percent of the sample—identified themselves as pure naturalists.

Of those who responded that they have a combination of beliefs, eleven described themselves as more naturalist than deist, four said they were equally naturalist and deist, and two said they were more deist than naturalist. Another eleven described themselves as having some component of theistic belief. Thus, thirty people in my poll said that they had some religious belief. (One of my respondents did not answer this part of the questionnaire.) But fewer than half of this group believed in a god who intervenes in the world.

My questionnaire asked seventeen different questions altogether, with space for optional comments, so it addressed many more issues than did earlier polls. One question I particularly wanted answered was whether evolutionists considered themselves monists or dualists. Monism, which is derived from the Greek word for "one" or "alone," assumes that all things studied in the natural sciences are guided by natural forces that can be studied through the methods of the natural sciences. Monists deny the existence of a supernatural realm that has any influence on the physical universe. Dualists, in contrast, allow for the existence of a supernatural entity. They believe that the universe is composed of a natural realm and a supernatural realm. Both theists and deists are dualists, because they believe in two domains: one consisting of the natural world and another the realm of God.

Monism is the default worldview of natural science. In science, an explanation has to be grounded in empirical evidence. In a slightly different take, for a statement to be considered a scientific explanation, it must be falsifiable—there has to be some kind of test that could be applied to the statement to prove it wrong. For example, the statement that the moon is made of cheese is a scientific statement because it can be falsified. Facts can be brought to bear on the claim (such as the density and geology of the moon derived from calculations of its orbit, observations through telescopes, and experiments on the rocks returned by the Apollo astronauts), and the claim can be rejected.

Another way to describe science is to say that it is based on the concept of skepticism. In science, the skeptic doubts the truth of a statement until evidence is available to support that statement. If untested claims exist in a scientific explanation, those claims point to areas that need more study and scrutiny.

Scientists believe that they can come closer and closer to something that can be described as "the truth" through observation, experimentation, and verification. They may never know if they have achieved the absolute truth—to the extent that such a thing can be defined. But if a statement has been tested so many times that there are no longer any reasonable grounds to suspect that further testing will reveal a discrepancy, scientists no longer refer to that statement as a theory or hypothesis. They call it a fact. Thus, it is a fact that Earth revolves around the sun, that human beings need oxygen to survive, and that biological evolution is responsible for the diversity of organisms that live on this planet. No serious scientist is testing these statements, because all of these statements have been verified beyond the point that additional tests are necessary.

In contrast, religions are worldviews based at least in part on the actions of a supernatural entity. Religious texts are not revised or made consistent with new knowledge, unlike the body of knowledge that constitutes science. Traditional religions do not encourage the discovery of new empirical knowledge unless it is done explicitly to verify the wisdom that is laid out in religious writings. The narratives of religion can grow richer, but they can never be falsified. Religious people may believe that "the truth" can be found from silent contemplation or from a private dialogue between a lone individual and a deity. But such knowledge is not grounded in the physical world. It is personal, unquantifiable, subjective, and internal. Therefore, it fails the test of the naturalist worldview, because truth for a religious person is not based on the pillars of naturalist knowledge: discovery, experimentation, and verification.

One of the great advantages of the naturalist worldview is that it serves as a basis for bringing people together under a common set of ground rules. Knowledge in science is public, not private, because it must be submitted to others for verification or falsification. A naturalist believes that the empirical truth is waiting to be discovered, and that we can all agree on the empirical truth so long as we believe in a few important criteria. Science can exist in any culture and any nation. It is a worldwide enterprise where people with radically different backgrounds can converge on the same truth. In an age when disagreements on issues of truth and opinion loom so large, the ability of naturalism to forge agreement on hard issues is one of its great attractions.

Naturalists do not recognize the supernatural, so it came as no surprise that monism was the worldview of the majority of evolutionary biologists who answered my poll. In response to the question of whether humans consist of only material properties, only spiritual properties, or both material and spiritual properties, 73 percent chose only material properties. Even more—88 percent—said that they reject the notion of immortality.

In my poll, I also asked the evolutionary biologists about their views on the relationship between evolution and religion. Here I have to admit that I had some expectations about what they would say. I knew, from previous polls, that the majority would not believe in God. So I expected them to say that religion and science are mutually exclusive. After all, religion makes many claims about the natural world. The Bible states that a great flood destroyed everything on earth, that the sun stood still, that Jesus was born of a virgin mother, and that the dead came back to life. Though some of these statements can be understood metaphorically, at least some are clearly meant to be taken literally, since much Christian theology rests on their veracity.

I was surprised by the answers I got. The majority of the evolutionary biologists (72 percent) said that religion is a social phenomenon that has developed with the biological evolution of our species. In other words, they see religion as a part of our culture. They do not necessarily see it conflicting with science.

This seems like social courtesy to me, not intellectual honesty. Evolutionists appear to be more concerned about remaining in the good graces of the public than they are about responsibly exploring the implications of their worldview. It may be possible to compartmentalize science and religion so that they seem not to conflict. But avoiding potential conflict between science and religion by not asking the tough questions sidesteps the confrontational spirit of scientific investigation. Claims made by authorities with the tacit expectation that they should go unchallenged out of reverence to those in power are precisely the kinds of claims I like to investigate and challenge. After all, the basic practice of science requires us to test all claims by the same criteria: observation, experimentation, and verification. If scientists are willing to rule out an entire domain of human life as exempt from their methods, how can they expect anyone to respect those methods? By trying to protect themselves from a public backlash against their overwhelmingly monist view-point, they undercut the very point they are trying to make.

+

At a wildlife preserve not far from my home in upstate New York, I sometimes see church groups of young kids on nature walks. So few Americans are interested in basic natural history that I find it hard to condemn this activity, even if a Sunday-school teacher is content to answer every question by saying, "Wow, God sure made some cool stuff." I taught my two children at a very young age how to observe nature closely—not to indoctrinate them into some sort of religious perspective but to teach them to see the world for what it is. The more people can observe the many bugs, amphibians, plants, and rock formations around them the better.

The kids on a church outing are essentially doing the work of natural theologians from a bygone era two hundred years ago. What I hope is that they undergo, in their own lifetimes, a progression similar to the one biology has undergone over the past two centuries. Several of the biologists I interviewed for my dissertation abandoned early religious beliefs when they became immersed in the natural world. One of them, John Bonner of Princeton University, told me, "The reason I decided one day that I didn't really want any religion at all at that age—well, I was maybe fourteen by this time—was that birds, sparrows outside my window, seem to be having a perfectly fine existence and are managing tremendously well. . . . I thought, 'They can do that without God,' so that's what made me decide that religion was not for me. [From] that moment on I really did not believe in God."

Perhaps the same thing will happen to those kids on nature walks. They may discover something that fascinates them and compels them to consult a scientific field guide. Maybe they will be motivated to study natural science in college. And by the time they reach college, they are likely to recognize the tensions between a dualist and a monist perspective. If they still believe what they were taught in Sunday school, they may find it increasingly difficult to maintain their dualist perspective as they become trained in science. As they perfect their own methods of discovery, experimentation, and verification, they may eventually admit that the monist naturalist worldview embraces their enthusiasm more harmoniously than does the religious worldview of their youth.

Naturalism is not for everyone. For many people, the hope of an eternal afterlife and a personal, internal relationship with God is at the core of their consciousness. Demands for scientific verification just stand in the way of their reasoning and dash many of their most coveted hopes. As far as I'm concerned, if a philosopher or theologian wants to interpret scientific data as divine, he or she has a right to do so. (Maybe they can write with a quill pen, too!) But when people deny empirical truth—

as when they scoff at evolution or contend that humans were created in more or less their current form just a few thousand years ago—they bring religion into conflict with science. I like to think that American society can be tolerant and honor intellectualism, but my hopes might be pipe dreams. If the fundamentalists have their way, we could enter a new era of intolerance and factionalism.

When I am teaching biology, I try to give students the facts and let them draw their own conclusions about what those facts imply. If I can lead them to ask important questions, the kinds of questions I was asking when I was a student, I consider my efforts successful.

I take the same approach when writing a song. I don't want to tell people what to think, but I want them to think. Sometimes that means pointing out hard truths and challenging them to examine their preconceptions with the hope of provoking more thoughtful dialogue. For our 2004 album, *The Empire Strikes First,* which was one of Bad Religion's most politically direct albums, I wrote a song called "God's Love" with the following chorus:

> *Tell me, where is the love?*
> *In a careless creation, when there's no "above."*
> *There's no justice, just a cause and a cure.*
> *And a bounty of suffering, it seems we all endure.*
> *And what I'm frightened of is that they call it God's love.*

Our decisions and actions reflect how we think about the world. Challenging authority and the dogmas that pervade our society does not make someone a crazed nihilist hell-bent on destruction. If done with an open mind, such challenges can contribute to informed social discourse and an open society. That's ultimately why it's so important to know about evolution: because it can change the way we think about ourselves and the world around us.

CHAPTER 3
THE FALSE IDOL OF NATURAL SELECTION

Natural selection does not act on anything, nor does it select (for or against), force, maximize, create, modify, shape, operate, drive, favor, maintain, push, or adjust. Natural selection does nothing.
—William B. Provine[1]

In high school, I was so eager to learn more about evolution that I applied for a volunteer job at the Natural History Museum of Los Angeles County. I had no experience with museum work, and my grades certainly didn't identify me as a promising junior scientist. But the museum's department of paleontology always had more field material than the staff could handle. They accepted my application and put me to work in the fossil-preparation laboratory.

I had to ride the bus an hour and a half each way to get from my mom's house in the Valley to the museum, but I didn't mind the hassle. The staff at the museum collected large quantities of rocks from the field, which were full of fossilized material. The fossil preparation lab had two full-time technicians, each with an elaborate bench full of tools and alien-looking fossils; a couple of university students with smaller desks; and a modest preparation station for volunteers. My job was to carefully chip away at the rock and discard the sandstone surrounding a fossil, using dental tools, toothbrushes, and a pneumatic device called a zip-scribe. As soon as I finished, the bone would be whisked away for study in another part of the building.

Most of this work was extremely tedious. It might take two hours to reveal a square inch of a bone or extract a tooth from its grainy sandstone matrix. My technique rapidly improved, but I also found that I wanted to learn something more about the animals whose fossils I was preparing. I knew there had to be a larger context within which these fossils were significant. I wanted to know what the fossils meant, not just what they were.

The placement of organisms into categories is known as taxonomy.[2] Many professions have this kind of expertise. An experienced brick mason knows more about the characteristics of bricks than any of us would imagine. He might be able to tell you what quarry the sand came from, how hot the fire was, and what kind of molding was used. His naming of all the different types of bricks is a little like the practice of taxonomy.

My work at the Natural History Museum was necessary for a beginner naturalist, because I needed to learn how to name and categorize organisms. I also took great pleasure in knowing the formal scientific names of plants and animals—it was a kind of secret knowledge that I had and most people didn't. But naming, labeling, and categorizing assumes a rigid order to things, and the really important and interesting things in life are not static; they continually change. What I was interested in was systematics—the relationship between fossils and modern-day organisms.

Have you ever met anyone with an encyclopedic knowledge of obscure rock bands? I knew a group of people in Los Angeles who spent their time browsing the used bins at record shops back in the days when music was recorded on vinyl (which is making a comeback these days, even though most kids have never heard anything other than compressed 128-kilobit-per-second digital recordings). Some of these people were so obsessed with obscure bands that they deserved the moniker "vinyl vermin." They collected lists of band names and knew all the rarest albums available. There could be an obscure garage band from England that released just five hundred copies of a single album. None of the rest of us would ever have heard of the band, but the vinyl vermin could tell you more about it than you ever wanted to know.

The problem with most vinyl vermin, I've found, is that they let their knowledge of trivia overwhelm their judgment. Despite their encyclopedic learning, I can't recall having a single discussion with them about whether any of the bands were actually any good. Maybe a band that released just five hundred copies of an album was an undiscovered gem, or maybe the music was so bad that no other record company would hire it to make another album. I never knew what most of the vinyl vermin thought about the qualities of musical groups or genres, because they never talked about anything other than trivial facts and statistics.

The lesson I learned from the vinyl vermin was that the most important thing about gathering information is what you do with it. The "secret language" of taxonomy might have made me feel special, but words applied to fossil species (or obscure records) didn't satisfy me. Taxonomy is a beautiful art. But without theory behind it, taxonomy amounts to words on a museum label. Even today, new species are being discovered and described at a remarkable rate, and each newly discovered species receives a unique official name. But what does the naming and ordering of species say about their relationship to other species and to us? I wanted wisdom, not just knowledge.[3]

I didn't realize it at the time, but my experiences at the Natural History Museum had interesting parallels with the history of fossil collecting in the years before Darwin. As early as the sixteenth century, naturalists in Europe and, to some extent, in other parts of the world began to collect fossils, exotic plants and animals, and other natural objects.[4] They would organize these objects into collections, which they often displayed in "cabinets of curiosities" that the public could view for a fee. Like old-time record stores, these places were repositories of obscure artifacts. Some of the largest collections became the cores of today's most famous natural history museums.

These early naturalists were dualists. They believed in an ordered, intelligently designed nature that was not constantly changing but rather was lying in wait for God's inquisitive children to discover. All species were assumed to be specially created by God. The obvious similarities between organisms therefore must be part of God's plan. But because these similarities were created by God rather than an intrinsic part of nature, early naturalists felt free to use their own schemes for arranging and naming the animals and plants in their collections. This caused utter confusion. It was as if all the vinyl vermin had created their own schemes for bands and musical genres without consulting one another or trying to adhere to a common classification. Without a proper taxonomy, God's design in nature might never be understood.

The problem of naming and organizing organisms was solved by the seventeenth-century Swedish botanist Carl Linnaeus. He devised a system that used two names for every species, in the same way that people generally have two names. The first name assigns an organism to a genus of broadly similar organisms. This is analogous to your first name. You might know a lot of people who have the same name as you. But your first name doesn't make you unique. In like fashion, Linnaeus's goal was to have the first name of a group of related species—called the "genus"—identify a type of organisms. For example, the genus *Canis* includes dogs, coyotes, wolves, and jackals. The second name (sometimes called the "trivial" name) identifies a unique subset of organisms within the larger genus. This subset is called the species. So the coyotes that roam the woods near my house in upstate New York are *Canis latrans.* By Linnaeus's system, humans are *Homo sapiens.* At the moment, we are the only species in our genus, *Homo,* though in the past, various other species in our genus existed, sometimes contemporaneously.

Linnaeus had an encyclopedic knowledge of the species that had been discovered, and he knew that they showed varying levels of anatomical similarity to one another. But he operated under the dualist assumption that nature was carefully planned by the Creator. His objective was to figure out God's scheme, not to question the supernatural wisdom of Creation. For example, in different parts of the world, different species seemed to live in very similar ways and serve very similar biological functions. But this did not require an explanation in Linnaeus's time. It was all part of God's plan. The socially acceptable form of intellectual pursuit was to show how various plants and animals conformed to an intelligently designed world. In the seventeenth through nineteenth centuries, there was very little opportunity to gain wisdom from the study of organisms.

✦

The explanation of evolution offered by Darwin and Wallace shattered the comfortable intellectual certainty of natural theology. They showed that slight variations in organisms made individuals more or less likely to survive in a given environment. If these traits were heritable, they could be passed along to offspring and would become more common in future generations. Darwin called this process "natural selection," because, in effect, nature "selected" from among the traits present in a population, in the same way that a plant or animal breeder selects organisms with advantageous traits in order to create offspring in which these traits are prominent. Natural selection offered a purely mechanical explanation for the incredible variations that previous biologists had attributed to God's handiwork: the camouflage of insects that resemble sticks or leaves, the beauty of flowers, the ferocity of predators, our own hairless skin, upright stance, and big brains.

Darwin knew that he had to make a strong case for natural selection. Many biologists, including his grandfather Erasmus Darwin, had previously speculated that evolution must have occurred because of the obvious similarities among living and extinct organisms. But the idea had not gained wide acceptance because no one could figure out how similar species might diverge from common ancestors. Natural selection provided a mechanism. It showed how populations of a single species could gradually acquire distinct traits over many generations without the intervention of a wise deity, simply through the profane activities of reproductive successes or failures. Natural selection was so important to Darwin's argument that the full title of his 1859 book describing his ideas is *On the Origin of Species by Means of Natural Selection, or the Preservation of Favoured Races in the Struggle for Life.* (By "races," Darwin was simply referring to groups of organisms that have particular traits rather than to the groups of humans that we might call "races.") Natural selection played a polarizing role in Darwin's worldview. It rooted his belief that natural theology was mistaken and that there was no "wisdom" in the design of nature.[5]

Origin of Species is essentially a book-length argument for natural selection. Darwin postulated that new traits arise more or less randomly in organisms, though he also speculated that some could arise through the repeated use or disuse of some biological part. These traits produce what Darwin called "differential survival and reproduction"—organisms live and die because of their traits. Those with a slightly more favorable complement of traits tend to have more offspring than those with less advantageous traits. Furthermore, Darwin recognized that organisms produce more offspring than the environment can support over the long term. Therefore, individuals engage in a "struggle for life," in the sense that all organisms have to compete over scarce resources. The result is a gradual, generation-by-generation increase in the suite of favorable traits. In this way, descendant populations come to differ from their ancestors. Darwin called this process "descent with modification," and he postulated that this process can lead to the formation of new species as populations of organisms gradually diversify.

Darwin's formulation of natural selection appealed to nineteenth-century Victorians. In Victorian England, the idea of a "struggle for life" fit well with the conditions of daily life. Infant mortality was much higher than it is today. Two of Darwin's ten children died when they were very young, and Darwin's daughter Annie died when she was ten years old, a tragedy that caused Darwin to give up the last vestiges of his religious belief.[6] For the industrial workers of the time, long hours of manual labor under excruciating and dangerous conditions for very low wages were commonplace. A struggle for existence seemed to describe the lives of many people at the time.

Natural selection resonated with nineteenth-century Europeans for other reasons. It struck many people as a process that explained and justified the enormous inequities in society. The weaker members of society would gradually falter and be eliminated—through poverty, hunger, and disease—while the stronger members would prosper and reproduce. Darwin's "struggle for life" was captured in a phrase coined by his contemporary, the philosopher Herbert Spencer—"the survival of the fittest." Even Darwin accepted this phrase as an excellent description of natural selection.

Natural selection also appealed to many scientists of the time because it seemed to bring order to the wild anarchy of the biological world. It provided a mechanism that seemed to have a logic and inevitability comparable to the laws of nature emerging from other sciences. By serving as the underlying mechanism behind "descent with modification," the branching tree of life seemed both inevitable and intelligible, combining extant species (those alive today) with extinct species known only as fossils.[7]

Natural selection even had a shadowy, theological appeal. It seemed to offer a direction or ultimate purpose to life. Over time, living things appeared to grow more complex. As new generations of organisms acquired new traits, they became progressively better adapted to their environments. What better evidence of a wisdom in nature preordained in the mind of God? Even for non-theists, the order created by natural selection might have seemed at least partially to compensate for the loss of God's oversight.

✦

Many explanations of evolution essentially end with a description of natural selection and its role in the diversification of species. These accounts provide a mechanism for evolution and assume that the proffered explanation covers it all. But over the years I have become more and more dissatisfied with natural selection as an explanation for all evolutionary change. And since I believe that dogma must be challenged wherever it is found—whether in religion, science, or music—I have spent time exploring the ideas of the iconoclasts who have examined natural selection critically. The result is a picture of evolution quite different from the standard textbook account.[8]

But before I look more closely at natural selection, I have to issue a blanket disclaimer. Whenever an evolutionary biologist identifies a problem with standard accounts of evolutionary theory, creationists tend to wave the statement around as evidence that evolution is fatally flawed or "a theory in crisis." That's ridiculous. As I've already pointed out, the occurrence of evolution is indisputable. The idea that God could have planted the entire fossil record in the earth as a way of testing the faith of believers is preposterous. I am not at all interested in leaving the door open for discussions with advocates of the modern "intelligent design" movement.

Still, all biologists know that the study of evolution is far from complete and may never be complete. Many fascinating questions surround the ways in which organisms evolve, the pace of evolutionary change over time, the influences of the environment on evolution, the relationship between the development of organisms and evolutionary change, and many other topics. Evolutionary biology remains a thriving field of science, which means that scientists continue to do research on unanswered questions.

The tendency of creationists to mischaracterize the statements of evolutionary biologists reveals their fundamental intellectual dishonesty. They spend their time attacking the ideas and words of biologists instead of offering verifiable ideas of their own. In particular, the group known as "intelligent design creationists" proclaims that their criticisms of evolution are based on scientific research. They say, for example, that some biological structures and functions are too complex to have evolved naturally and must reflect the design input of a deity. But intelligent design creationists have not provided a single scientific finding that backs up these claims. Their research, such as it is, consists largely of wishful thinking. If they have made an impact, it's been the opposite of what they intended: when a new biological structure or mechanism is discovered, evolutionary biologists tend to get busy and look for evidence consistent with evolutionary theory.[9] Creationists rely on a God-of-the-gaps approach, where areas of scientific uncertainty are equated with the mysteries of God's actions. Yet this approach can be very damaging to their cause, because the general course of science is to fill in gaps over time. The paradox comes from the recognition that more discovery brings more gaps. When a new finding occurs in a gap, two new gaps are created on either side of the finding, creating new holes in the continuum of facts. Thus, the more data, the more mystery. Intelligent design creationists, insistent that God exists in these gaps in our knowledge, learn less and less about God's actions and intentions as more and more discoveries are made. Intelligent design advocates should stop now, since they're getting further behind with each new discovery.

The real agenda of intelligent design creationists is not scientific. They intend to substitute an overtly theistic worldview for the naturalist and monist worldview of evolution. Their attack on evolution is a means to an end. For example, a major proponent of intelligent design creationism has been the Center for Science and Culture, which is sponsored by the Discovery Institute a conservative Seattle-based think tank. The objective of the center, as described in an internal document leaked to the public in 1999, is "to defeat scientific materialism and its destructive moral, cultural and political legacies."[10] The idea is to use intelligent design creationism as a "wedge" to separate science from its allegiance to "atheistic naturalism." Although the center has not produced a single shred of data to support intelligent design, it has sponsored conferences, supported popular books and articles, organized lectures, and worked with conservative politicians on proposals to teach religion in science classes. These efforts have met with great success. Proposals to force biology teachers to discuss creationism in public schools are regularly introduced in local school boards and state legislatures, though most of the proposals have been defeated. But the campaign continues unabated.

Sometimes creationists argue that educators should be forced to "teach the controversy."[11] But there is no scientific controversy, just a social controversy. My PhD adviser, Will Provine, acknowledges this social controversy by inviting creationists to talk with the students in his non-majors course on evolution. For many of his students, the dialogue Will has with the creationists is very refreshing. Will allows his students to incorporate creationist thinking into their term papers. Many of his creationist students recognize a great similarity in their thinking with the pre-Darwinian natural theologians. They can still earn high grades in his class, even if they cannot wrest themselves from the creationist way of thinking. A fairly sizable minority of his students, however, have changed their views on creationism after taking his course. For the first time, they have been able to think carefully about how Darwin changed the scientific worldview. They have come to understand that intelligent design creationism is not science, and it should not be treated as such.

But back to natural selection. Darwin did not know the biological mechanisms responsible for the inheritance of traits between generations. Then, during the first decade of the twentieth century, scientists began developing the discipline of genetics, which provided a missing ingredient in Darwin's theory. Geneticists ascribed the inheritance of biological traits to what they called genes that were passed from parents to offspring through the sperm and egg cell. For decades, they did not know what these genes were or exactly where they were located in cells. But they determined, through experimental breeding studies, that traits pass from generation to generation in particular ways, even if the exact nature of "genes" remained mysterious.

The development of genetics gave rise to one of the most important scientific developments of the twentieth century. In the late 1930s and early 1940s, several geneticists in the United States and Europe combined ideas from genetics, evolution, and mathematics to create what is called the modern synthesis (or the modern synthetic theory of evolution).[12] The modern synthesis is a beautiful theory. It combines insights from many fields of science to create a quantifiable and predictable account of evolutionary change. Through the rigor of its mathematics and the scope of its explanatory power, it quickly became the reigning paradigm in evolutionary biology.

In part, the modern synthesis is natural selection on steroids. The modern synthesis sees natural selection as an active force that constantly monitors the traits of the organisms in a population. These traits are assumed to be in a one-to-one correspondence with the genes in those organisms. Random changes in genes provide the variant traits on which natural selection acts. Natural selection is therefore an evolutionary overseer. It picks and chooses among the different traits exhibited by the members of a population.

The modern synthesis relies heavily on the notion of the "fitness" of organisms. Fitness is a theoretical measure of how well an organism is adapted to its environment. It can be measured, in a rough way, by counting the number of viable offspring that an individual has. If another member of the same population leaves a greater number of descendants than do others, then that individual is said to have higher fitness. Fitness at a local gymnasium or spa depends on how much an individual exercises; fitness in nature depends on how well an individual reproduces.

According to the mathematical formalism adopted by the modern synthesis, species occupy hilltops in a sort of fitness landscape, where high points in the landscape (peaks) represent regions of greater fitness. When the genes and therefore the traits of a population change, the population moves away from the peak toward a valley in the theoretical landscape. If it ends up crossing the valley and rises to another hilltop, the population persists. But if genetic or environmental changes force a population into a fitness valley, it will be ruthlessly eliminated by natural selection, along with all of its traits and genes.[13]

Before I get to the problems with this view, it's important to note that the modern synthesis has had some unfortunate consequences. First, it has placed a tremendous emphasis on genes and on DNA as a source of biological traits. We are said to live in a genomic age, as if everything in biology can be reduced to statements about genes. Yet DNA is just one part of our biological machinery and is unable to do anything on its own. It would be equally shortsighted to give the central role of punk rock to the lyrics of its songs, ignoring the musicians and the punk fans who form the collective environment of the punk subculture.

The overemphasis on DNA has produced some serious distortions in conceptions of genetics and in how people think about biology in general. Undergraduates are routinely told to focus more on genetics courses and labs than on field studies. As a result, students never get a good sense of variation among wild populations. Maybe their professors think that they are more likely to find jobs in laboratories than in fieldwork. But there is also an undercurrent of thinking in most biology departments that genetics is simply more important than other types of biological investigations. For example, most medical schools require at least one year of genetics for their incoming students. Yet they don't require a single semester of comparative anatomy, dissection, or biological fieldwork.

The public's perception of DNA's importance is also skewed. Many people think that studies of DNA will produce cures to common human diseases. But the promises have been mostly hype, despite a few notable advances (such as the production of human insulin in bacteria). The sequencing of the human genome, for example, was supposed to trigger a golden age of personalized medicine and biomedical breakthroughs. That hasn't happened, at least not yet (and there are reasons to suspect that it may never happen in the way people expect). Genetics research was supposed to reveal the simplicity lying at the heart of biology, just as natural selection was supposed to reveal the simplicity at the heart of evolution. Instead, the more we learn about genetics, the more complicated it gets. Much of the DNA in humans has no known effect on our bodies. Other large parts of the DNA are "control" regions that turn on, turn off, or modulate the activity of genes, depending on factors in the environment. The sections of the DNA considered "genes" are relatively sparse, yet even these sections of DNA interact with one another and with the environment in countless and poorly understood ways. People can send DNA samples away to be tested to see if they might have an increased risk of disease. Yet the results they receive are difficult to interpret at best and often needlessly alarming. So far the genomic age has been largely a bust.

To be fair, the modern synthesis was a major advance in the understanding of evolution, and I have learned a lot from studying it. But many biologists have observed things that do not fit well with the modern synthesis. Some have tried to twist their observations to fit the paradigm rather than questioning widely accepted views. But in doing so, they have obscured the true richness and diversity of biology. The modern synthesis has created a caricature of biology that has robbed evolution of much of its capacity to serve as the foundation of the naturalist worldview.

+

One of the first serious cracks in the modern synthesis came in the 1960s, when molecular biologists began to study the constituents of proteins in organisms. These proteins turned out to be much more variable than expected. Proteins of the same species showed large amounts of variation, even though they functioned more or less the same. And none of the varieties of proteins seemed to have any selective advantage over any other one.[14]

This didn't make sense. According to the modern synthesis, natural selection should have removed all but the most "fit" variations of proteins from the species. If proteins were so variable, natural selection couldn't be maintaining an iron grip on evolutionary change.

Actually, this finding wasn't that surprising to field biologists. Anyone who has spent much time observing organisms in their natural habitats knows that organisms engage in many counterproductive behaviors that would never have persisted if natural selection were strong. When I was an undergraduate at UCLA, I took a field-oriented class away from campus to study leaf-cutting ants in the tropical deciduous forests of central Mexico. The species I studied (from the genus *Atta*) builds trails on the forest floor. You might think that ants aren't big enough to make a trail in the soil, but if you look closely, you can see shallow depressions, clear of leaf litter, forming thoroughfares between three and ten inches wide through the forest. The trails continue to the bases of trees, where the ants march up the trees, into the canopy, and out onto the leaves, where they begin their harvest of tiny leaf fragments. Their large mandibles slice through the waxy coating of tropical leaves. Each ant then hoists a disk-shaped fragment of vegetation overhead and returns along the trail system to the nest, which might be three hundred or four hundred feet away. Inside their underground nests are repositories where the workers place and further disarticulate the leaf fragments. Essentially, these repositories are underground fungus gardens. Fungi grow on decaying leaves, and the colony subsists by feeding on them.

This all seems like a well-oiled operation that could rival the most efficient human corporate factory. But when I began to study the process, I immediately encountered a tremendous amount of inefficiency. In many instances, unladen workers were heading home empty-handed, and even many laden workers were walking the wrong way on the trail or roaming far out of their way. Furthermore, the "ant highway" extended much farther from the nest than seemed necessary. Of the four plant species they harvested, three occurred in abundance only fifteen feet from their nest, yet their nightly excursions took them hundreds of feet away to harvest different plants. What a waste of energy!

This kind of chaotic phenomenon is commonplace in nature. Faithful adherents of the modern synthesis try to explain such chaos in terms of natural selection, but their arguments quickly become highly convoluted. They might tell you that the inefficiency and chaos are only illusory. Perhaps the chaos evident in the behavior of leaf-cutter ants enables the colony to explore more widely and exploit new resources. "In the long run," they might say, "all ant colonies will have to compete for limited resources, at which time natural selection will be ruthless in eliminating the less efficient ones."

But I suspect that the *Atta* ants demonstrate something else, and something very important, about life. In the here and now, there is plenty in nature that has nothing to do with natural selection. Much of what we see in nature actually represents the wastefulness of abundance, where natural selection, as it is traditionally described in textbooks, has very little effectiveness in whittling down the natural anarchic exuberance of life.

Another vivid reminder of natural selection's lack of explanatory power occurs every autumn around my home in upstate New York. October brings an abundance of fall color as the leaves on deciduous trees turn bright red, yellow, and orange. As night temperatures drop and sun exposure wanes, biochemical changes take place in the leaves of trees. The conventional neo-Darwinian explanation is that natural selection favors trees that drop their leaves to prevent damage to delicate leafy tissue, whereas the twigs and branches are hardy enough to withstand winter's harshness. This explanation is fine, but it says nothing about the most conspicuous trait of deciduous trees in autumn—their color!

The actual explanation involves the biochemistry of trees. When the sun falls on leaves, it creates an automatic biochemical response within tiny specialized organelles inside their cells. They produce a photosynthetic pigment called chlorophyll, which is bright green to our eyes. Chlorophyll enables the tree to make sugars for the plant's consumption. As the photoperiod diminishes in the fall, less chlorophyll is produced, and other photosynthetic pigments begin to reveal themselves. Most plants make numerous types of pigments, but when sunlight is strong, chlorophyll dominates the other pigments and masks their effects. In fact, some of the pigments, such as the red pigments, are amplified by the short, warm days of fall.

The important point to recognize is that the most conspicuous and brilliant trait of deciduous trees has little to do with natural selection. Instead, it is produced as a simple by-product of sunlight abundance (insolation) and fall temperatures. It reminds me that some of the most enjoyable aspects of life are largely accidental.

The inadequacy of natural selection as an explanatory principle is especially apparent when we look at the species with which we are most familiar: ourselves. Many biologists use the neo-Darwinian approach to explain our traits. They postulate that natural selection has been responsible for our most cherished characteristics—our upright stance, our physical features, our big brains, our language. But a careful examination of these traits has turned up far more questions than answers.

Undisputed evidence for natural selection in humans exists for very few traits. For example, natural selection has commonly been invoked as the primary cause for human skin color in different parts of the world. Roughly speaking, people with ancestors who lived closer to the equator have darker skin than people with ancestors who lived at higher latitudes. This is due to melanin, a protein pigment found in many vertebrate animals. Human skin cells produce melanin as a reaction to sunlight, and some individuals make more of it than others. I, for instance, get lots of freckles when the sun shines, but my blotches of melanin don't do much to protect me from the sun, and I get terribly burned very easily.

Two evolutionary factors seem to be responsible for my skin color and that of other humans.[15] In areas with intense sun, dark coloring protects the skin from damage. Africans living in equatorial regions who are albinos and do not have melanin in their skin are at severe risk for skin cancer, whereas light-colored skin (without much melanin) in someone living in Norway poses far less danger. In fact, lacking melanin if sunlight is less intense might be an advantage, because it enables the body to absorb more sunlight. This sunlight helps to produce vitamin D in the body, which is necessary to avoid a variety of diseases. And the DNA of northern Europeans and Asians has skin color variants that seem to have been favored over evolutionary time, supporting the idea that natural selection has caused the skin of populations in northern areas to lighten.

Yet even with skin color, which many textbooks use as a classic example of natural selection in humans, all kinds of anomalies exist. The indigenous people of Tasmania lived for many thousands of years on their isolated island. Yet when Europeans first landed on Tasmania in 1772, the skin of the Tasmanians was still as dark as that of their African forebears, even though Tasmania is about as far away from the equator as Italy. Similarly, humans first migrated from Asia up and down the length of North and South America more than ten thousand years ago, which should have given their skin plenty of time to adjust to differences of insolation. Yet when Columbus sailed to the Americas in 1492, he found dark-skinned people in tropical latitudes, and their skin tones were not much different from those of native peoples in northern latitudes, such as the Inuits or "Indian" tribes of Canada. Insolation, therefore, is not such a good example of natural selection, after all. And of what use are freckles, anyway?

I suspect that skin color, like many traits in humans and possibly other mammals, is more a product of human sexual selection than of natural selection as it is traditionally defined.[16] If people preferentially choose mates with particular skin colors, natural selection is not a factor at all. Instead, traits are perpetuated simply because people find them attractive. The same process could apply to height, body size, body shape, and other human characteristics. All sorts of cultural factors affect perceptions of human traits, which means that culture can have a big influence on human evolution. It could be that culture has its own set of rules, and the genes just follow along. What's more, the same process could occur in many other species. Birds, for instance, constantly engage in displays of their plumage and singing to attract the females with the highest fitness. If choice is involved, there are many factors that might play a role in causing a female to be attracted to certain behavioral and physical traits. Sometimes the seemingly fittest males just don't do it for a choosy female. And without successful reproduction, an organism's fitness is zero.

For many years, anthropologists have sought to attribute other physical characteristics of human groups to natural selection. For example, they have speculated that the epicanthic fold of the eyes in eastern Asians evolved to protect their Ice Age ancestors from the glare of the sun off blinding snowfields. Or they have said that the short, squat stature of pygmies was an adaptation to the equatorial heat of Africa (while ignoring the fact that tall, lean Africans lived just a few hundred miles away). In actuality, almost all of these speculations have turned out to be unjustified and unverifiable just-so stories. Recent studies have shown that most of our physical features have changed more or less randomly as modern humans spread from our eastern African homeland into the rest of the world.[17]

The point I want to make is that using natural selection to account for everything can quickly become convoluted and tedious. In the previous chapter, I gave a plausible explanation for the evolutionary changes in *Tiktaalik*'s highly developed front legs. But who really knows if *Tiktaalik* developed stronger and longer legs to spend more time out of water or to search for prey? What if these remote ancestors had females that simply preferred to mate with males who had more robust limbs? Is that natural selection in the wild? The actual evolutionary mechanism could have been quite different from conventional views of natural selection.

It's important to remember that arguments over evolutionary mechanisms often depend on the deeply held convictions of evolutionary biologists—after all, biologists are people, too! Almost everyone is reluctant to let go of deeply held convictions. Many biologists, in particular, are still attracted to teleological accounts of evolution. Teleology is the idea that all activities tend toward the achievement of some sort of goal. "Everything as to its purpose" is the undercurrent of teleology, a philosophical worldview championed by the ancient Greeks. From this perspective, the goal of a species is to reach the hilltops in the fitness landscape I described earlier. If a species does not attain this ideal, then it must be fixed by natural selection or eliminated lest it suffer a suboptimal life. Many biologists assume that most of the variation seen in a given species is maladaptive and that only a small subset of a population attains optimality. But this is the opposite of what we see in nature. Variation dominates in all species. If all traits were optimized by an ever-watchful natural selection, variation would quickly be eliminated in nature. For example, plant breeders have removed almost all of the variation in most of the fruits and vegetables we eat, which has resulted in predictable, market-ready (and often tasteless) food. But natural populations have none of the uniformity of conventional products.

The moment a biologist opens the discussion to optimality, all of the seemingly maladaptive, anarchic, and random features of life stand as stark counterexamples. Biologists committed to the supremacy of natural selection seem to want to replace God the designer with Nature the designer. And, in that case, why can't Nature, in its wisdom, simply be a manifestation of God? In that regard, teleology (and its handmaiden optimality) plays right into the hands of intelligent design creationists. An optimum is an ideal, an abstraction, just like the theological explanation of God's purpose for everything in the universe. If we aren't performing to our potential, we are committing a sin against God, or a sin against natural selection.

+

A minority of biologists have been examining mechanisms other than natural selection that may be at work in evolution, and they have found an abundance of candidates.

First, as mentioned in the previous chapter, DNA is not the only carrier of information in the sperm and egg cells that form the link between generations. Egg and sperm cells contain lots of other molecules—including proteins, structural compounds, and snippets of a DNA-like molecule called RNA—that can change the traits of an offspring. In addition, DNA molecules are festooned with other molecules that control whether particular sections of DNA are active or not. The experiences a person has during his or her life can alter these molecules on the DNA in egg and sperm, causing changes in an offspring that are not a consequence of changes in DNA sequences, which is anathema to true believers in the modern synthesis. For instance, mice fed diets high in methylated food have higher levels of methyl molecules on their DNA, and these alterations can be passed on to offspring. So not all hereditary information flows through the letters of DNA molecules, and an organism's experiences can affect its inheritance.[18]

Biologists have long known that organisms with more or less the same DNA sequences can develop quite different traits depending on the environment in which they grow up.[19] You can see this for yourself by observing the trees and vines in a forest. Some trees, particularly in tropical forests, have a marked difference in the shape of their leaves from the lowest to the highest branches. Even though all the leaves of a tree have identical DNA, they show vastly different color patterns or shapes, depending on the environment where they develop. In addition, other environmental factors important to trees, such as temperature, water abundance, and pests, vary vertically in the forest, and these factors also can affect the development of leaves.

Another simple example requires a trip to the zoo. Flamingos are pink because of their diet. In the wild, they eat a variety of crustaceans, from which they derive pigments (called carotenoids). If they find other foods in the wild that are low in carotenoids, they will develop white plumage. Indeed, most of the bold coloration of birds is derived from pigments in their diets. The most brilliantly colored individuals have the most successful matings. In the parlance of evolutionary biology, they have the highest fitness. Thus, a bird's plumage can change depending on the local abundance or scarcity of food. In this way, the environment can play a drastically significant role in determining a bird's reproductive success, regardless of the genes it carries.

These examples bring to mind an uncomfortable confession: I have hideous feet. Ever since I was a kid, I've played a lot of sports, and my feet have been jarred, twisted, and bruised in ways that they probably shouldn't have been. The result: calluses as thick as a pachyderm's skin, bone spurs, and asymmetry. Skin cells throughout the body all carry the same genes. But if you expose certain areas of your skin to abrasion, you get calluses in those areas. Bone and muscle are similarly plastic. They are highly sensitive to physical stress. If even slight amounts of force are applied over long periods, the phenotype can be drastically altered. Just look at "before" and "after" pictures of someone who has had braces. The jaw and teeth can be reshaped, with a concomitant change in the chewing muscles. Yet no genes were altered in the transfiguration.

In short, humans make choices, and those choices, not natural selection, can determine the traits that are expressed. Human children do not just inherit DNA from their parents. They inherit an environment in which human beings talk to each other, live in houses, use computers, and sometimes go to church. This social environment causes some varieties of traits to be perpetuated in much greater abundance than others.[20]

In general, many observations show that evolution does not look like a well-oiled machine designed and controlled by natural selection. I tend to think of evolution as more like a waterfall. We can understand the properties of water under various conditions and the factors that might interact with gravity to change the water's direction. We can study the topography of a stream and identify characteristics like water depth, maximum discharge, sediment load, or organic debris that affect other properties of the waterfall. But a waterfall is always changing. Sometimes portions of it are frozen, other times it is barely trickling. In spring it is rushing with abrasive sediments that change the shape of the stream channel and erode the knickpoint (the point where the water begins to fall). Any sequence of snapshots we take to study the intricacies of the waterfall will show slight differences over time. But all the pictures together, like growth stages of a human individual throughout his or her life, refer to a single thing that is constantly changing with the environmental conditions that surround it. Organisms with their traits are like the water flow-ing over the lip of the cascade. Some molecules will fall straight down, while others will be diverted by rocks and the molecules around them. Sometimes air molecules will be a factor. With high winds, portions of the falling stream will dissociate into mist. The entire system is in constant motion and is very difficult to describe thoroughly.

Each biological entity consists of immense populations of molecules, cells, tissues, organs, individuals, and species that are all interacting simultaneously, at every instant of life. Nothing in the universe could be more complicated to describe. The individuals existing at any given time are the result of an incomprehensible number of causes, both historical and immediate. We can identify some of those causes, but it is very difficult to be confident that one type of cause is more important than another in establishing the pattern of evolution. Evolution is the result of all causes acting in concert.

I haven't even mentioned yet one of the most important factors influencing the course and outcomes of evolution: sheer, blind chance. Many organisms live or die not because of natural selection but because of random, impossible-to-predict events. Perhaps a meteorite strikes the Earth and wipes out a significant fraction of the species on the planet. Or an egg cell is deposited in a barren location and can't develop. Or a pod of killer whales happens upon a school of fish and only some get eaten. Or rats carrying the plague board ships and travel from the Mideast to medieval Europe. Who knows how many vagaries have been involved in life's history? Evolution creates a tremendous number of opportunities through its endless, profligate creativity. Which of these opportunities is realized—and which tragedies counterbalance the opportunities—may be largely a matter of chance.

Chance is so whimsical that it is hard to either analyze or analogize. But anyone who questions the importance of chance in biology should spend some time thinking about its role in his or her own life.

✦

I've always felt that I was tremendously lucky during my adolescence. Things easily could have turned out differently for me had I gotten in trouble with the law or gotten addicted to drugs or alcohol. Almost every night, instead of studying, my friends and I would head to Hollywood to be where the action was. Drugs, sex, and trouble could be had on any given night, and the police and punkers fought a running battle over control of the streets. I was lucky to get through that period of my life without doing great harm to myself or others. Many of my teenage peers weren't so lucky.

The Los Angeles metropolitan area covers about five thousand square miles, and in the early 1980s the city's punkers were dispersed throughout the region. But the hangout that embodied the coherence of the L.A. punk scene was Oki Dog, a rundown hot-dog stand at the corner of Santa Monica Boulevard and Vista. Actually, "hot dog" can't begin to describe an Oki Dog. It's a pile of rubbery wieners, American "nacho" cheese, chili, and peppers wrapped in a massive tortilla. I ate hundreds of them while idling away endless hours with my punk entourage. Oki Dog was appealing for everything a sixteen-year-old punker could want: greasy food and easy girls.

This part of Santa Monica Boulevard was known for its hookers, zonked-out drug addicts, cruising gays, and all varieties of punkers. On any night of the week, from ten to a hundred leather-clad, spike-haired night-owls would congregate at Oki Dog, both to get something to eat and for the social intercourse. Punk celebrities like Darby Crash, Axel G. Reese (singer of the Gears), and Belinda Carlisle frequented the place, bumming cigarettes and fries off other people. The place was full of misfits who came to criticize your fashion sense or geographic integrity: "Oh, you're from the Valley? What kind of loser are you?" Or "I've seen punk wristbands like that on sale at Posers on Sunset Boulevard."

Our group was normally lighthearted and jocular. Besides me, it usually consisted of Jay Bentley, Greg Hetson (who was on his way to becoming punk royalty with the Circle Jerks), Arnel the boat man, Alex, male nurse Roger, Peter Finestone (who had replaced Jay Ziskrout as our drummer the previous year), Kenny "No-chin," smirking Kevin and his "wife" Becka, and the "ten high" club of teenage drunks—Lisa, Laurie, and Shannon. On most nights, we sat around telling jokes, laughing at other people, and discussing upcoming shows. We tended to sit on the parking blocks on the east side of the building rather than the more visible picnic tables in front, which were under bright lights and were the showcase spots for socialites of a higher status. Even preadolescent punkers were welcome to hang out with us. The members of a band called Mad Society made a name for themselves about that time because of their eleven-year-old singer, Stevie Metz, who sang a song about their Asian heritage: "When I was a boy in Vietnam, we were hit by napalm, napalm, napalm, we were hit by napalm."

Every part of the property was used. The Hollywood punks would assemble at the picnic tables and on the corner sidewalk. The Orange County punkers usually sprawled next to us in the east-side parking area. The beach punks scoured the sidewalks on either side of the building, often skateboarding and taunting passers-by. The girls at Oki Dog were mostly free agents. None of them was loyal to any one group. The girls liked our humor and felt comfortable around us. But they mostly wanted to pair up with older guys who had more experience than we did and lived on their own in some squat or fleabag apartment nearby. Our being from the Valley didn't play well in Hollywood. Heroin abusers and established musicians always pulled with greater force than did our schadenfreude and good-natured meanderings.

At that time, I was mostly blinded by the desire for sex, but Oki Dog socializing turned out to be an important activity. Status was achieved in those days by something that today we would call networking. Bad Religion had some success with a 1981 EP we released on our label, Epitaph Records.[21] In 1982, we released out first album, *How Could Hell Be Any Worse?*, and it got lots of attention, so plenty of people knew about our band. But some of our best contacts were made in the parking lot at Oki Dog, which led to important gigs that came from invitations by friends in better-known bands. In particular, Bad Religion's connection to the Circle Jerks, one of L.A.'s most popular and influential punk bands at the time, was fostered by Greg Hetson's friendliness and willingness to participate in our jocund musings at Oki Dog. His band members talked about us and played one of our demo tapes during a guest appearance on *Rodney on the Roq*, a Sunday-night radio show on KROQ hosted by Rodney Bingenheimer, which was an important outlet for new bands at the time. It was the first time I ever heard my voice on the radio. After that, Rodney added us to his weekly rotation, and Greg Hetson started making guest appearances with us at every Bad Religion show.

The Adolescents were one of the most popular bands from Orange County at the time. Our friendship sprouted at Oki Dog after shows when they had gigs in L.A. Eventually, they asked us to play shows with them. I remember one time they visited my mom's house during one of Bad Religion's rehearsal afternoons. I felt as though we were hosting true royalty. I serenaded them with piano renditions of their most popular songs, and we all had a great laugh. We're still great friends.

The mix of people who gathered at Oki Dog also created the potential for trouble. On a particular night in the spring of 1982, when I was a high school senior, we were sitting at Oki Dog when we heard some yelling along the boulevard. "That's right, I'm gay. What about it?" Some skate punks had been taunting a gay couple that was headed home from a club. In a flash, the man and his partner were beset by the punkers. The skaters chased the couple into the parking lot and knocked them down. One of the men fell backward and hit his head on one of the parking blocks. I still remember the sickening sound of his head making a loud thwack on the cement, and then the other skaters started in on his friend as he tried to run away. As other people at Oki Dog gathered around to watch and then to help the injured man, Jay and I agreed that we had seen enough and promptly jumped in his truck to head home. Later, news reports said that the gay man died from head injuries after a fight with punk rockers on Santa Monica Boulevard.

That was a pivotal moment in L.A. punk history. It reinforced a reputation for violence that was growing rapidly in the punk community. Skinheads and other groups with a propensity for violence had begun coming to many shows. The police exacerbated the situation by cracking down on the bands and their fans, who, in turn, lashed out at the police.

The terrible death at Oki Dog capped a day of other harsh events still prominent in my memory. Earlier in the evening, a girl in my high school had a party to celebrate her parents' departure for a short vacation, and I heard that a punk band was going to be playing there. This sounded strange to me, because I knew there were no other local bands in the Valley. In fact, the only PA system in the area belonged to Brett, and it was permanently stationed in my mom's garage, where we practiced every day. Earlier that evening, Jay and I had shown up at the party. We found it to be crowded with new punkers, many of whom had just shaved their heads or dyed their hair that afternoon. The music blasting from the stereo was punk standards of the day, but we had no idea what band was setting up its gear on the backyard patio. "Some new band, I guess. They're coming out of the woodwork these days," I told Jay.

But Jay had more knowledge of the skate-and-surf scene, and he quickly recognized some of the people who were coalescing around the patio. They were members of the legendary Dogtown skaters from Venice. It was all completely alien to me. While they were inventing the modern culture of skateboarding, I was just a grade-school kid in Wisconsin. Now I was face-to-face with them at a party in my own neighborhood out in the seldom-visited hinterlands of L.A. One of them told Jay and me that we'd gotten there just in time to hear Jim Muir's brother sing. Jim Muir was an inventive skateboarder and surfer who founded Dogtown Skates in 1976, when he was eighteen. His brother, Mike, had become lead singer for a band called Suicidal Tendencies, which was playing one of its first gigs that night.[22]

After the band had played a few songs, I ventured outside, because Peter Finestone had arrived with some familiar faces from the Oki Dog crowd. Suddenly, I noticed some other punkers in the front yard being taunted by a group of dark figures backlit by a streetlamp. Apparently, the punk party had aroused the attention of another party of high-schoolers, which was taking place just a few houses away. When one of them moved closer to yell some inarticulate insult, I recognized him as someone from my class who also happened to be on the football team. In fact, the eight people around him were all members of the El Camino Real High School football team. "This is going to be interesting," I thought. Every day at school I worried about being picked on by these guys, since I was one of just a few punks in my school. Now, for once, I was surrounded by punkers, and the football players were outnumbered.

While the jocks sent a minion back to their party to retrieve some baseball bats and trash-can lids, one of the punkers ran inside to tell the band to stop playing because there were some Valley kooks outside talking shit. As the music from the backyard abruptly ceased, the jock standing in the front yard yelled, "Go have your party somewhere else, faggots." Then, much to my horror, I heard one of them yell, "Graffin, is that you? I see you, you fucking fag. This party must be your idea. Tell your Devo pussies to hit the road."

At about that moment, the Venice entourage and about twenty-five other punkers came pouring out of the front door and walked directly up to the jocks from my high school. "Who are you calling Devo pussies, motherfucker?" asked Jay Adams, one of the Dogtown originals who later that night was involved in the assault outside Oki Dog. The jocks, now aware that their numbers were insufficient, stood motionless and feigned bravery. Another skate punk walked up to one of the jocks and yanked a baseball bat away from him. In an instant, the jocks began to retreat as their own weapons were used against them. They scurried into the house where their party was being held, while the punkers followed. Sounds of crashes and screams and banging echoed through the neighborhood. Some of the jocks locked themselves in closets; others were chased down the streets. Aside from a few landed punches and kicks, the jocks weren't badly injured. But the house where they were having their party was trashed. Doors were ripped off the hinges, windows were broken, upholstery was slashed, and I'm sure that the place would have been set aflame if someone had had a blowtorch.

As I sat back and watched the chaos, I felt a tremendous sense of comradery and admiration for the fearlessness of the Venice skate punks, who were demolishing the pride of the narrow-minded jocks who made my life at school a living hell. At the same time, I knew that I was in deep shit. I was the only punker the jocks identified by name. The Venice boys were from someplace else and would be gone by the end of the evening. Jay and Brett had already dropped out of high school, so they didn't have to deal with the jocks anymore. Bad Religion was synonymous with punk in the West Valley, and at our high school, I was synonymous with Bad Religion. I would surely be a target of retribution when I showed up at school on Monday.

The punk party quickly disbanded after the razing of the jocks' house, and we all filed into vehicles bound for Oki Dog as the first sirens headed our way. As I got into a car with Mike Muir, Tony Alva, and Jay Adams, I heard one of the jocks yell, "See you in school on Monday, Graffin!" I thought I was a dead man and that school would be my final resting place.

I can't remember what compelled me to get into the car with

the Dogtown skaters. All three were drunk or high or both. I remember heading over Cahuenga Pass on the freeway while one of the passengers sat on the sill of an open window, swung open the door, and pissed on the car next to us—at sixty miles per hour! During the drive, I remember explaining my greatest worry. "They're going to kill me at school on Monday." Mike Muir tried to comfort me. "Don't worry, dude. We'll put a fucking bomb in their mailbox if they start any shit!" But I knew that the Venice punks were never coming to the West Valley again and that I was going to be on my own.

It turned out to be a shitty Friday night—no music, no gigs, just mean street violence. I remember thinking that this kind of life wasn't for me. I was always happiest when I was performing or attending a gig. Now all I could think about was getting assaulted Monday at school with the full fury of the football team and their Friday-night embarrassment.

Monday morning, I somberly put on my jeans and T-shirt and headed to school. The first attack came between first and second period, while I was walking in the hallway. A fierce blow struck the back of my neck. It was one of the jocks from the party on Friday night, and a couple of his friends were there, ready to join in. But, apparently, someone had told the school security that the jocks were going to kick my ass, because before they could descend on me, the school bouncers intervened and separated me and my assailant. I remember the rage in my classmate's eyes as he was held back by security. "I'm gonna kill you, motherfucker. They chased me with a chain."

One of the jocks came up to me. He was the quarterback of the team, and a friend of mine. I told him what had happened, and he told me how scared and angry the team was. When I explained that I had nothing to do with the event except having the bad luck of being there, he promised to try and calm down his teammates. Meanwhile, rumors were flying around the school. People said that I was a member of a punk gang that was going to take on the jocks. But really it was just me, alone and with no backup.

In my math class just before lunch, the school security officer entered the room and asked, "Is Greg Graffin present?" Everyone turned to look at me, and I raised my hand. The officer led me into the hallway, where police officers were waiting. They explained that they were going to place handcuffs on me and take me to the police station. They had called me into the hall so that the other students couldn't see. But just as my hands were cuffed behind my back, the bell rang and their plan to be discreet was blown. Students poured into the hallway, headed for lunch, just in time to see the police escorting me down the long corridor, outside along the walkway, and into the police car parked at the curb. It was tremendously humiliating. But at the same time the police did me a huge favor. With that single "perp walk," the police had turned me into someone to be feared, even though everyone I have ever befriended in high school or elsewhere knows that I've never done anything to warrant such a reputation. "Graffin must have killed that guy in Hollywood this weekend" was a report from my friend at school who heard the whispered rumors.

I was never formally arrested. The police explained that they were investigating the property damage at the house on Friday night, and I was the only person named in the plaintiff's report. On the drive to the station, I told them that I didn't know any of the people at the party and that they weren't from around here. Then they lectured me about how the West Valley is a community and those punk friends I hang out with in Hollywood are no-goods who care little about our neighborhoods or communities. I'm sure that I showed no appreciation for their advice. But, in retrospect, I realize that the police had the good sense to recognize that I wasn't some sort of kingpin and that I wasn't interested in leading a life of violence. I was back in class by the final period of school that day. The entire student body thought that I must be legit, now that I had been "arrested." I wasn't bothered by jocks anymore, and Bad Religion had a great rehearsal at my house later that afternoon.

✦

If I had any degree of emotional maturity in 1982, I would have recognized that my desire to be where the action was could have been disastrous. I wasn't smart enough to avoid trouble. I got into a car full of of drunk daredevils and arrived unscathed at Oki Dog later that evening only through good fortune. I could have gotten the beating of my life from the jocks if my encounter with the police hadn't earned me some respect. But I was smart enough to understand one thing at the time. If punk was turning into a life of fights, property damage, fear of retribution, and run-ins with the cops, then I wanted nothing to do with it.

I was never attracted to the illegal and dangerous aspects of punk. For me, the thrill was always in the intellectual challenge to authority inherent in the music that I and others were writing. But that challenge was being interpreted in violent terms. Fans and the police would fight outside shows, and lots of gigs were shut down. Most of the bands were not violent, but there was no stopping the flow of converts to the punk lifestyle. Bullies who just a couple of years earlier were shouting demeaning insults at punkers now cut their surfer locks and came to shows to drink beer and kick ass. It was a confusing time for many of the punk bands. They recognized that their aggressive music was being used to justify the fighting, even though their lyrics often spoke out against such narrow-minded behavior. Partly as a result, many of the bands began to break up or morph into different kinds of musical acts. It's no coincidence that the less-violent "glam" metal scene began about this time in Los Angeles, fueled in no small way by musicians who grew their short hair long and adopted outrageous personas.

Bad Religion had some important gigs in the summer of 1982, including a show at Florentine Gardens with TSOL and two nights at the Whisky a Go Go with Social Distortion. All of our shows came off without any incidents, and they gave me some welcome spending money for the summer before college. I was reassured by my friend Tony Cadena from the Adolescents, who sang a song in those days called "Rip It Up":

> Have you had enough violence?
> Just to kill makes no sense.
> We're not the background for your stupid fights.
> Get out of the darkness, it's time to unite.
> Do you think you're tough when you rip it up?

I was confused. I didn't want to sing for an audience that was just going to start fighting. But I felt no affinity with the glam rock scene that was beginning to take root as a response to the violence. Instead, Brett and I started to write an album that would forever remain our one departure from punk. It was called

Into the Unknown, and we released it on Epitaph Records in 1983. The songs on it reveal that neither of us wanted anything to do with the violent subculture of punk. Yet we also were obviously unaware of how it would be perceived by the people who enjoyed our music. The album sounds nothing like punk. It's more like Jethro Tull and Pink Floyd getting together with REM (though REM had just released their first album and we had never heard of them). To this day, I have mixed feelings about the album. As with any studio project, I learned a lot about making records. But I also learned that, as a musician, you cannot depart too drastically from an established style and still satisfy your audience. Looking back on it, I can't say that the album is very good. It was more of an unconscious rebellion against fellow punks who threw in the towel in a different way. Whatever was fresh and original about Bad Religion was gone by 1983 and the release of *Into the Unknown.* The album also essentially broke up the band, since Jay walked out of a rehearsal for the album and didn't come back to the band until 1986. Given the trauma of the preceding year, it is understandable why none of us wanted to be a part of the punk scene. I was soon off to college, and Brett was off to learn about the music business. It would be two long years before I entered the recording studio again.

Except for my growing interest in science and music, my adolescence was a careless, wandering meander. The things I did, the places I hung out, and the people with whom I associated were dangerous, and, if things had gone slightly differently, I might have been expelled from school or arrested, or become addicted. As it turned out, I came through it relatively unscathed. The outcome of my adult life has been mostly disconnected from the mistakes of my youth. I can only attribute that to luck.

There's no simple explanation for good fortune. When someone wins the lottery, we don't ask him to make sense of it—we simply say he must be lucky. We can study the way he lives his life, and the behaviors that led to his habit of buying lottery tickets, but that doesn't help us understand how he picked the right number. By the same token, it isn't necessary to attribute the diversity of life to a single factor—natural selection.

Species venture through time, and their "success," however you want to define it, is a product of good fortune as much as good genes. A species interacts with a constantly changing world, creates new associations with each turnover of the population, and deals with immediate environmental obstacles. A species has little to guide it through ever-changing ecological mazes, because its traits are no more than gifts from defunct ancestors and are not guaranteed to function smoothly under new conditions. Until the time of a species' extinction, there is very little that is predictable about the course it takes. Some species will rapidly flourish with huge populations, while others will wander aimlessly through ecospace with low numbers of individuals and few prospects for future success.

Punk rock was nearly derailed by its association with violence. It could have disappeared entirely by 1983 had not a few bands persisted through the doldrums of the mid-1980s. It stayed alive, like a Cretaceous mammal, sheltered in out-of-the-way places, until conditions were suited for its eventual flourishing.

CHAPTER 4
THE FALSE IDOL OF ATHEISM

The source of man's unhappiness is his ignorance of Nature.
The [way] he clings to blind opinions imbibed in his infancy . . .
dooms him to continual error.

—Baron d'Holbach[1]

My fifth-grade teacher, Mr. Gorsky, a strict former U.S. Army sergeant, didn't appreciate anyone talking out of turn and was intolerant of smart-ass comments. One day he gave us a homework assignment to memorize the planets of the solar system, and the next day he called on me to recite the names of the planets in sequence.

"Pluto, Neptune . . ." I began.

"From the sun outward, please," he said.

This confused me. I had learned the names of the planets from a poster that drew the solar system in three dimensions, so Pluto was closest and Mercury was farthest away. I decided to ignore him and do it my way.

"Neptune, Uranus . . ." I said.

"No, it's YOUR-i-nus," he said.

At that point, I was being embarrassed by a teacher who didn't appreciate my reverse-order learning of the planets and kept interrupting me. Recognizing a setup too good to ignore, I said, "But, Mr. Gorsky, there are no Klingons around YOUR-i-nus; they're only found around Ur-A-nus!"

The entire fifth-grade classroom burst into laughter, but Mr. Gorsky's face immediately clouded with anger. He marched to my desk, grabbed my arm, and dragged me into the hallway. I figured that I was headed to the principal's office, but instead he decided to give me a speech right there. He told me that I was an embarrassment to myself. He wagged his finger at my nose and said, "Don't you realize, young man, that those kids were laughing at you, not with you?"

I immediately knew that he was wrong. The class loved it when I cracked jokes. Some of my best friends were good at comebacks, too, and we laughed about our fifth-grade witticisms at recess. Mr. Gorsky was trying to convince me of something that had no basis in fact. I knew he had no way of verifying his opinion that I was a laughingstock. It seemed obvious to me that he was saying what he wanted to be true, not what actually was true. Maybe this kind of discipline worked in the military, but even as a fifth grader I recognized that, without any proof, one opinion was no more valid than another. In fact, Mr. Gorsky's misrepresentations worked against him. I had plenty of teachers who simply laid down the law and didn't try to justify themselves: "Okay, the rule in here is simple, if you speak out of turn, you will be sent to the principal's office." These teachers may not have been very inspiring, but they were taken more seriously than Mr. Gorsky was. They didn't confuse their formal authority with the expectation that we students would see the logic in following their rules. They believed that the most effective authority was one that invited no scrutiny.

But Mr. Gorsky had all the power and control, and I didn't. It wouldn't do any good to fight the power structure in that school. And I didn't want my mom and dad to learn about my continuing disrespect to my teachers. Even though I yearned to do otherwise, I capitulated to him that day in the hallway. "Yes, sir, I guess you're right. They're laughing at me. And it's better to learn the planets from the sun outward." But silently I was thinking, "You're wrong, Mr. Gorsky, and someday I'll prove it."

I tell this story not to demonstrate my insolence as a child—you already knew that—but to recall some of the difficult situations that kids have to deal with when they are confronted by small-minded restrictions on their thinking. Children are smart, independent people who often aren't able to recognize when they've gone too far in the logical extension of their beliefs. Their enthusiasm can be nipped in the bud by adults, and kids are usually powerless when confronted with these demands. "You can't really believe that" or "How could you have done such a thing?" are common statements of parents. Besides shaming the kids, these kinds of statements also indoctrinate children by giving them a rough outline of the social expectations in their community. As kids acquire their own individuality, somehow they need to reconcile their actions and beliefs with what adults tell them through the perceptual maze of childhood experience. And they do not have a very large base of experience with which to do that.

So far in this book, I've been telling stories from my adolescence to show how music and science led me to the naturalist perspective that I use to make sense of the world. But in this chapter and the next one, I need to break with that chronology to examine two important subjects. The first is the contribution that powerful childhood experiences make to who we become as adults. The second is the way tragedy shapes and distorts our worldviews. No one can escape these influences on their lives. Yet people can interpret them very differently, depending on the intellectual framework they use to make sense of the world.

+

As I've said, I've never believed in God, which technically makes me an atheist (since the prefix "a" means "not" or "without"). But I have problems with the word "atheism." It defines what someone is not rather than what someone is. It would be like calling me an a-instrumentalist for Bad Religion rather than the band's singer. Defining yourself as against something says very little about what you are for.

That's my biggest objection to the wave of atheist books and Web sites that have come out in the past few years. Simply put, atheism does not offer a constructive worldview. Embracing atheism can, of course, radically alter one's worldview, which I believe is the primary factor responsible for the popularity of books by the "four horsemen" (Richard Dawkins, Christopher Hitchens, Sam Harris, and Daniel Dennett).[2] But atheism forms only a portion of the naturalist perspective, and a negative one at that. Calling someone an atheist does not offer a way to build socially meaningful relations and institutions. It narrows perspectives rather than broadening them. Nor does atheism necessarily imply a sure path to knowledge, which I believe can be found only through the study of nature, of life, and of human societies. It is not clear how to put atheism into action in our society. Atheism as a term seems only to make people angry, whether adherents or detractors.

There's another problem with defining yourself in opposition to a particular worldview. Because atheism is defined through negation, it's never clear which meaning of "God" one opposes. Some believers revere an interventionist God who regularly influences physical events. Others believe that God rarely if ever exerts any influence over human affairs. Some people believe that God is evident in nature, while others believe that the existence of God can be revealed only through supernatural revelation. Many people believe in more than one god or even in a vaguely defined "spirituality" that does not require the existence of a specific god or gods.

Atheists can hold a similarly broad range of views. For example, an atheist can be someone who has no belief in gods because of a lack of interest in the subject, or someone who believes that gods do not exist. These latter atheists may be proponents of a specific philosophical position that acknowledges the possibility that proof of God's existence might materialize someday. Some people believe that not enough evidence exists to prove or disprove the existence of God, and therefore call themselves agnostic. But if they believe that not enough evidence exists to prove the existence of God, they meet at least one criterion of atheism, and their attitudes about most things may be indistinguishable from those who call themselves atheists. Similarly, people who consider themselves "spiritual" may be de facto atheists, even if they don't call themselves that.

Many people share my antipathy toward the word "atheism," though their reasons are very different from mine. Many Americans, even those who would never admit to being prejudiced against other groups, exhibit a strong and irrational prejudice against atheists. According to a 2004 poll, Americans rate atheists below Muslims, recent immigrants, gays and lesbians, and other minority groups in "sharing their vision of American society."[3] According to this poll, atheists are the minority group that average Americans are least likely to want their children to marry.

This fear of the word "atheism" prevents many people from calling themselves that, even if they have little or no belief in God. When Americans are asked in polls whether they are atheists, just a few percent say yes (the numbers vary from poll to poll, depending on the exact wording used and populations sampled). But far more people say that they are not religious, in which case the extent of their belief in God is at least open to question.[4]

Why are people so fearful of an intellectual position that most know very little about? Their visceral, unreflective attitude suggests that they are acting on beliefs acquired during childhood. But why are these beliefs so strong and so resistant to change? Are all childhood beliefs essentially immutable, or can we resist the influence of things we learn before we even realize we are learning?

When I visited Richard Dawkins, we talked about the possibility that natural selection has favored the suggestibility of humans at young ages. He said:

> Perhaps child brains are shaped by genetic natural selection to follow a rule of thumb that says, "Believe whatever your parents tell you." You can easily see why in general that rule of thumb would have genetic survival value. The world is a dangerous place. Children don't have time to discover by trial and error; it's too dangerous to learn such things as "don't swim in the river because there are crocodiles." You just have to believe what your parents tell you.

> [Religious myths] can continue down through the gen-
> erations because the child brain is set up with that rule of
> thumb: believe what your parents tell you.[5]

From this perspective, atheism is an idea that needs to be ad-
vocated and spread widely to compete with the very successful
religious ideas (which Dawkins labels "memes") perpetuated by
parents.[6]

This remains a controversial idea. Natural selection may have
very little to do with the adoption of ideas. Still, we all recognize
that some experiences and stories from our childhood stay with
us our whole lives. Their influence on our behavior is rarely scrutinized when we are older, which, I believe, is why atheism is so
foreign to most people. Almost all kids are told that God is the
ultimate authority. "He created the universe and you as well." It's
very rare for parents to say to their kids, as my parents did to me
and I did to my own children, "It's okay to question many of the
things I tell you. I'm not sure where we come from. Go find out
for yourself. Just stay out of traffic." If more parents said this,
more people would challenge authority and there would be more
tolerance for diverse worldviews.

I favor the idea that children are very susceptible to parents'
influence, because my parents did have a very great influence on
me. But their influence was not on my relation with God. It was
on my relation with music.

+

Some of my very earliest memories are of hearing my parents singing little melodies in the other room. There was always humming

or whistling going on if my dad was nearby. I always got a good feeling from my dad's songs, and that feeling returns nowadays when we get together during holidays or other special occasions. For my entire life, my dad has been liable to break into song at any moment. Music just comes naturally to him, like to a warbler or mockingbird.

The words of his songs also had a lasting impact on me. To this day, I can remember the lyrics to silly ditties he sang, small snippets from popular tunes of a bygone era: *"Bongo bongo bongo, I don't want to leave the Congo . . ."* or *"You're in the army, Mr. Jones, no more private telephones, you had your breakfast in bed before but you won't have it there anymore."* My mom had her own blend of favorites. She sang in community choirs, and each year they performed seasonal concerts. One of my earliest memories is of seeing her in the chorus of *The Pirates of Penzance.* Around the house, during this time, she would sing the part of the lead character: *"I am the very model of a modern Major-General, I've information vegetable, animal, and mineral, I know the kings of England and I quote the fights historical, from Marathon to Waterloo, in order categorical."*

Today, I find myself humming, singing, and whistling around the house or out in the yard, and I'm sure it affects my children the same way it affected me. There's no deep meaning to whatever song pops into my head, but it's obviously affected by my mood. When I'm in a good mood, I sing, for no apparent reason other than a feeling that comes over me, and I'm often in a great mood when the kids are around. When they were younger, they would try to stump me by saying a word, and then I would have to come up with a lyric from an actual song and sing it for them. So, if they said "pizza," I would sing "Crew Slut" by Frank Zappa:

"Hey, I'll buy you a pizza," or, if they said "candy," I would sing "Candy" by Iggy Pop: *"Candy Candy Candy I can't let you go."* On and on our game would go until they were exhausted. I usually could figure out a lyric to every word their young minds could come up with. Sometimes I had to cheat and use some of my own lyrics. I have written nearly 150 songs and collaborated on scores of others with Brett. All of the lyrics I perform onstage are hardwired into my memory. I don't have to study before I head out to play a concert. I simply hear the music and the words roll off my tongue.

I was born in 1964, so I remember virtually nothing about the 1960s. But the rock and roll of that decade and of the 1950s still had a huge influence on me. My dad listened to Johnny Cash recordings; my mom loved pop radio. I have fond memories of Dad cranking his cheap stereo to the point of distorting the paper-thin speakers. *"We got married in a fever, hotter than a pepper sprout, we've been talking 'bout Jackson, ever since the fire went out, I'm goin' to Jackson!"*

Racine, where Dad stayed after my parents' divorce, is about twenty miles south of Milwaukee, so I spent many hours in the backseat of cars driving between the two cities to spend time with parents, grandparents, and friends. It made me feel worldly to live in two cities. Many of my best friends in Racine grew up on the same block as I did, which made Racine seem like a small town to me. Then, in Milwaukee, I could do big-city things like going to sporting events or other happenings. It was perfect—I could feel cool with my big-city school crew, and then on weekends I had my neighborhood friends around my dad's house.

Most of the driving was Dad's duty. He picked my brother and me up from school every Friday and brought us back to

Mom's house every Sunday. The radio was a constant companion in his car. Dad never bought a car that cost more than $500. Some ended up lasting a few years while others broke down right away, but they all had one thing in common: really crappy radios. I remember a Plymouth Fury where Dad had to hit the dashboard every few minutes to get rid of the static, even on stations with the strongest signal.

In Dad's car, since there was no FM, we mostly listened to AM talk radio, radio dramas from the old days, and, in the summer, baseball—Brewers, Cubs, or White Sox games. I mostly tuned out in the backseat. When the weather permitted, I rolled down the window and stuck my head out to sing my own songs. This really bothered my brother, who would yell at me to stop singing. But Dad tolerated my singing just fine and never discouraged me.

Cruising around in Mom's car was a different experience. Mom's cars were always of a higher caliber than Dad's. The one I remember best was a 1969 Buick LeSabre that had speakers mounted on the rear deck. We listened to WOKY in Milwaukee, all-hit radio. From the time I was in second grade in 1972, I could sing every Top 10 radio hit every week. I would turn my entire body around on the backseat and face the cars behind us (there were no seatbelt laws in those days). I was the perfect height so I could lay my head directly on top of the speaker and sing along with the great pop singers of the 1970s: Stevie Wonder, Elton John, Simon and Garfunkel, Paul McCartney, Don McLean, James Taylor, and all the rest.

In the front seat, Mom always sang along. But she didn't like to sing melodies the way the pop stars did. She liked to sing harmonies to the pop songs we heard on the radio. Sometimes it annoyed me, because I wanted her to sing along with me, and all

I could figure out was the melodies. But I think her harmonizing had a profound effect on the way I ended up hearing music. Today I can't write or sing a melody without also hearing the harmonies in my head, and most of the Bad Religion repertoire features two-, three-, and four-part harmonies.

My dad had an excellent collection of classical, opera, and jazz albums, with a few pop albums mixed in. Whenever he played the Weavers, Pete Seeger, or the Beatles, he expected a sing-along. We especially liked to listen to live albums, complete with the cheering audience between songs. Listening to those albums was a family affair that we all cherished.

Mom's stereo sounded better than Dad's, but her record collection was not as diverse or large. If Mom played an album, it was most often Gilbert and Sullivan or some opera sung in a foreign language. Mom always sang along with the music, but she usually listened while she was cleaning house or cooking. I was content to listen from the other room rather than join in the household chores.

In 1972, however, she bought an album that combined her operatic sensibilities with Vietnam-era rock culture: *Jesus Christ Superstar*. I remember her singing the radio hit "I Don't Know How to Love Him" from that album. Although the song depicts Mary Magdalene's feelings for Jesus, it must have resonated with the new demographic of independent women who had left their husbands. I remember lying on my stomach on the floor of Mom's house, with the stereo speakers only a foot away from each of my ears, and learning every word and every note of *Jesus Christ Superstar*. My favorite character was Judas, always skeptical and caring. I was too young to know much about music. But I remember thinking that I wanted to sing like Judas. That album taught me not only about great musicianship, singing talent, and studio production, but also the basic story of the New Testament. What a bonus! I didn't have to read the Bible to get the gist of Jesus' life.

The first album I ever owned was *The Jackson 5's Greatest Hits,* which my mom bought for me in the second grade. I identified with Michael because he was so gregarious and just a few years older than me. I played that record to death. Once I went to a Milwaukee Bucks basketball game with my dad and one of his friends, and when we went out for pizza after the game we heard a Jackson 5 song. "This is my favorite band," I said. "Holy cri—" said my dad's friend. "That girl sure has a high voice." Having my favorite singer pegged as a girl was an affront to me, but I chalked it up to another example of the older generation not keeping up with pop culture.

I never told anyone about my love of pop music after I became a punker. But the pop sensibility in my songwriting and singing was still there, despite my best efforts to hide it.

In fifth grade, I was introduced to progressive rock. One of my best friends in grade school, Jeff Shimeta, had a next-door neighbor named Lester Savage, who was in high school. Lester was a prog-head who played for us every obscure LP coming out of England that made its way to the local record shops. I listened enthusiastically to Lester's albums by bands like Return to Forever, John McLaughlin and the Mahavishnu Orchestra, Steve Hillage, and Hatfield and the North, but I didn't much appreciate the music. What I appreciated was Jeff's setup for listening to music. He was an only child, so his parents had let him turn their basement into a music dungeon. Black lights and psychedelic neon-colored fishnets hung from the ceiling. The carpet consisted of square-foot shag samples from the local rug stores, glued piece by piece to the concrete floor. The walls were painted dark purple and black and were adorned with posters by Roger Dean. A strobe light and lava lamp lay on the component rack. His amplifier and turntable were top-of-the-line, and his speakers were medium-sized cabinets with mysterious black grilles covering the woofers and tweeters. I didn't mind listening to Lester's strange progressive music so long as we could play it loud at Jeff's place, and turn on the black lights.

These days, I see similar behavior in my children. The musical tastes of my daughter and her friends change weekly with whatever is being played on the national Top 10 countdown on pop radio. But instead of using high-end audio equipment and descending into a specially built listening environment, they congregate around a tiny laptop and go through the playlists on various Web sites or on iTunes. The music coming from the laptop's coin-sized speakers is weak and tinny, and the vocals are highly processed and most often drive the speakers to distortion. I can't understand why they don't go upstairs to my studio and listen to their songs on a high-end stereo. But the girls couldn't care less about sound quality. They dance, sing along, and discuss the songs and the singers. My take is that they're developing a sense of belonging. It doesn't really matter how it sounds. What is more important is: Who is making the music? What does each song represent? Who else likes this music? The music gives the girls a connection to one another and to the world at large. They believe that a culture exists, beyond their immediate circle of friends, in which they would be welcome.

When I listened to music with Jeff, it was the first time that I had a sense of belonging to a world larger than our Wisconsin neighborhoods. With Lester's influence, we believed that other people would accept us because we liked progressive rock. This feeling of connection may manifest itself in unpredictable ways. People can be attracted to different kinds of music or to other stimuli entirely. Where a feeling of connection comes from depends on a finite but incalculable combination of factors. I don't judge musical genres, and I don't put people down for listening to music that I don't understand—they are just taking different paths of social and intellectual development. Indeed, becoming connected to music with Jeff was an important development for me.

Luckily, Lester also had some more mainstream prog-rock albums by bands like ELP, Yes, Pink Floyd, Genesis, and Jethro Tull, which were more song-oriented and catchy than traditional jazz fusion. Probably most important for me, the vocalists were all top-notch. I recognized at an early age that a group could consist of superb musicians but what really brought a song to life was the delivery of the vocalist.

Jeff and I were both in the school choir. We had an excellent music director, Jayne Perkins, who let us sing current pop songs in our biannual concerts. Every morning before rehearsal, Mrs. Perkins let us spend some time listening to rock albums, and the kids usually brought in the most popular albums of the time: Elton John, Led Zeppelin, the Beatles, the Jackson 5, Queen. But Jeff and I had a secret weapon that let us scoff at the ignorance of our benighted peers. We knew about Todd Rundgren. When Jeff and I decided to buy Mrs. Perkins a birthday gift, Lester highly recommended *Another Live* by Todd Rundgren's new band Utopia. We had never heard of the album, and neither had Mrs. Perkins when we presented it to her. But we were so impressed by our own good taste in modern music that we bought a copy for ourselves. That album had a huge effect on my future musical style.

This was about the time when I figured out how to record

LPs onto my cassette recorder. Whenever it was convenient, I borrowed an album from Jeff and recorded its highlights onto a blank cassette. I was able to copy a bunch of ELP and Pink Floyd, and they were perfect additions to my artistic-mood music collection. I also recorded FM radio with my Sony. In those days, FM radio played hit music, even though the stations billed themselves as "album rock." I avoided recording Led Zeppelin songs, and Kiss songs, and Ted Nugent songs, and Thin Lizzy songs, because for some reason I associated these songs with a more mainstream type of kid whom I thought of as a "burnout." But every once in a while they would play "Roundabout" by Yes, or the shortened versions of "Bohemian Rhapsody" by Queen, or "Dogs" by Pink Floyd, and I would push the red "record" button and edit on the fly to keep the DJ's commentary off my mix tape. I had no means of buying a good music collection, so I was happy to have a few tapes of great music.

I never heard Todd Rundgren on FM stations, so I believed I knew something that DJs didn't. I thought that Jeff and I had stumbled onto a great secret that only a select few were privileged to know. Todd Rundgren was, to us, an invisible god, so powerful and yet so ethereal that the average music fan couldn't hope to criticize him.

Many years later, I had the opportunity to work with Todd as a producer for one of our Bad Religion records. In person, he was disheveled, brash, and impatient. The other band members really disliked his style of production, but since I was so familiar with his body of work, I saw our collaboration as a creative step forward. The album we produced, *The New America*, is still one of my favorites. But the process of working with Todd every day in the studio took its toll on my childhood idealism. Instead of being a god, he became simply a colleague. What I worshiped about him prior to the project started to diminish as I saw more of his negative personality traits. His talent, however, was shocking. He's the best guitarist I've ever worked with, and his song crafting is phenomenal. All in all, I got something more important from Todd than my childhood adulation could provide. I gained practical inspiration and a friend—and what can be more meaningful?

I never had formal music lessons, but I spent a lot of time at my mom's piano, learning to play chords by myself. My mom had a spinet piano, and most afternoons after school, if I wasn't listening to music, I was playing at the piano and dreaming of performing music. The first songs I ever learned to play were Todd Rundgren songs. Singing was natural to me, so I often made up words and sang songs with simple chord progressions. I wrote my first song in about the third grade, but the lyrics could have been written by a kindergartner toying with the key of C: *"Bye bye bye bye, seen her blue eyes, I seen her blue eyes, and I couldn't recognize, I taught her how to dance, but she didn't take a chance, I taught her how to dance."*

Such piano ramblings were a source of companionship and comfort in the quiet, latchkey-kid world of a divorced household. I didn't hang out with my older brother much, and my mom was busy at work, so my after-school activities were centered around the stereo and the piano. Basically, I was whiling away my time until the weekend, when I could go back to Dad's house in Racine and play sports with my friends.

✦

These early childhood experiences had a profound effect on my life. I'm sure that some of the chord progressions I've written into

Bad Religion songs were ones I played on my mom's spinet after school in Milwaukee. And certainly the lack of religion in my upbringing affected my worldview. I never had to rebel against a wrathful god or risk everlasting fire and brimstone.

Much of what we believe about some of life's most important questions—where we came from, how we should act, what we should believe, our purpose in the universe—undoubtedly originates in our childhood experiences. Stories, traditions, and behavior patterns of those around us, not to mention the consequences of traumatic events, often can't be overturned by what we learn later in life, except possibly through vigilant dedication to change. Maybe some of these beliefs are nearly permanent because they are incorporated in neural connections established in our developing brains during the early years of life, which would explain why people find it difficult to accept challenges to their most strongly held convictions.[7]

Most children are compelled to obey and do what they are told. Their worldview is laid out for them by their parents and other adults who expect them to accept that worldview without reservations. Rebelling against these parental injunctions is extremely difficult and sometimes impossible, especially when parental expectations express themselves so obliquely that we don't recognize their source. For children in a religious family, in particular, religion forms the internally consistent logic of their lives, and scripture erases any desire to be skeptical. As Romans 14:23 states: "He who has doubts is condemned."

Religious beliefs may also be related to a person's temperament. Research has shown that believers tend to be more conscientious and less open to new experiences, whereas nonbelievers tend to be more open to new experiences and more socially active.[8] In addition, a major event can lock us into particular ways of thinking or feeling, which may be why so many religions feature developmental milestones during adolescence, like bar mitzvahs or confirmations. Religious education in the form of Sunday school, parochial school, or prayer groups can reinforce religious norms. With all of these influences on our religious outlook, it's amazing that anybody ever changes their mind.

But plenty of people do.[9] Many young adults raised in strongly religious households eventually give up their belief in God. Almost all of them began by asking questions: Why can't women become priests? How could God influence events in the physical world? Why does God allow so much human suffering? A gap begins to widen between their worldviews and those of the religious believers around them. They may begin to question the premises on which their faith previously was built. Many of these individuals suffer rejection from their parents and friends, yet they persist in their search for the truth. Many find their way to what is essentially a naturalist worldview, in which they use the human power of discovery to learn what exists in the world outside their own heads and what is a figment of their imagination. To me, these are stories of great courage and persistence in the face of daunting opposition.

Acknowledging that other people or social institutions have led you astray is very difficult, especially if those people are your close family members. Most people who reject scientific narratives are desperate to retain their worldviews, and new information is very threatening. Accepting new information means finding a new way of reasoning about, and possibly even understanding, your own behavior. It can mean breaking with friends and family, or a constant struggle to hide the truth from those around you.

Some people compromise with a naturalist perspective by saying, "I'm not religious, but I'm still spiritual." I've always had a hard time figuring out exactly what that means. When I ask them, they tend to say, "You wouldn't understand what I believe," or "It's my own personal form of belief." In other words, they are not willing to submit their beliefs to scrutiny. Some of these people may be uncomfortable with traditional religions. They may feel that no one religion represents their views. They might have been raised in a religious household but do not accept much of what they learned. But they still crave supernatural explanations for their ultimate questions.

Some people claim that being spiritual means only that "I believe in things greater than myself." These believers could be either monist or dualist. In fact, I myself believe in a lot of things that are greater than myself. But I am a staunch monist who feels perfectly content being part of a chaotic, unpredictable unfolding of organic events that began billions of years ago.

Others define spirituality in terms of awe or wonder at what they might call the "mysteries" of the natural world. I love mysteries and don't recoil from things about which I feel unsure. On the contrary, I'm motivated to learn more about such things, confident that new truths will ease mystery and add to my sense of connection with the natural world. And just because something is mysterious doesn't mean we should automatically jump to the conclusion that science cannot explain it. In fact, some of these mysteries actually are not so mysterious to those who have investigated them. Scientists may not have been very good at communicating their explanations to the general public, but if scientific explanations were more widely known, fewer people might claim to be "spiritual."

People have so many differing beliefs and convictions that it is hard to specify exactly how many people believe in God. As I mentioned earlier, the number of people who are willing to call themselves "atheists" in the United States is very low. But many other people are atheists in everything but name. While they may be unwilling to label themselves as such, their beliefs nonetheless may have nothing to do with God of Scripture and may therefore be very offensive to traditionally pious people. According to some polls, more than 20 percent of Americans consider themselves atheists, agnostics, spiritualists, New Agers, believers in eastern religions, or some other type of nontheists.[10] Together, this group is as large as any denomination in America.[11]

The number of people who do not believe in God is much higher in other countries: 31 percent in Britain, 48 percent in France, and 54 percent in Norway.[12] Polls have found that as many as 82 percent of Swedes do not believe in a "personal god." Counting the atheists, agnostics, and nonbelievers in all countries, between 500 million and 750 million people on the planet are nontheists. And in many countries, including the United States, rates of nonbelief are climbing. This is a large fraction of the human population, roughly equivalent to the fractions believing in the world's major faiths. I suspect that many of these people have some sort of a naturalistic worldview, though no general survey data exist.

Countries with a high percentage of nonbelievers are among the freest, most stable, best-educated, and healthiest nations on earth. When nations are ranked according to a human-development index, which measures such factors as life expectancy, literacy rates, and educational attainment, the five highest-ranked countries—Norway, Sweden, Australia, Canada, and the Netherlands—all have high degrees of nonbelief. Of the fifty countries at the bottom of the index, all are intensely religious. The nations with the highest homicide rates tend to be more religious; those with the greatest levels of gender equality are least religious. These associations say nothing about whether atheism leads to positive societal indicators or the other way around. But the idea that atheists are somehow less moral, honest, or trustworthy has been disproven by study after study.

+

I don't promote atheism in my songs or when I teach undergraduates. During my lectures about Charles Darwin, for example, I barely mention Darwin's decisive reasons for abandoning theism. Far more important is his theorizing about biological phenomena. The focus of students' attention at the introductory level, where I teach, should be on the processes and interrelationships found in nature. The debate over whether species are specially created by a deity has only a secondary significance, and there simply isn't time to discuss it in introductory biology class.

I am, however, happy to share my opinions, if anyone asks. Darwin showed that human beings are a part of nature. Though we have unique characteristics, we are no more exalted or advanced than other species. (After all, all species are unique.) When students understand the logic of this conclusion, it tends to shake their confidence in humanity's privileged position as God's favorite creation. Or at least they are encouraged to think more carefully about their beliefs and suppositions.

When I'm writing songs, I try not to do what Mr. Gorsky did to me. I don't force adherence to a particular worldview. I simply suggest that there may be a way of looking at things that is more productive than traditional religious outlooks. Maybe it's a reflection of the experience I had while listening to *Jesus Christ Superstar,* but I have always believed that songs can change people in deeply significant ways—not only how they feel but how they think. If, as some scientists maintain, music draws on a different kind of learning than "book learning," then singing about ideas rather than feelings may be a way to tap into these other modes of learning.[13] Because of this—and, of course, because of the way my parents treated me—I've always adamantly supported my children if they want to listen to their iPods instead of reading an assignment. I realize that it's usually a ploy not to have to do homework, but I believe strongly nonetheless that they are learning something of value by listening to music.

When I began to write lyrics of my own, I tended more to ask questions than make absolute statements. I have always believed that making too strong a claim in a song degrades its artistic integrity. For instance, there's a place in "God Song" from the Bad Religion album *Against the Grain* where I got a bit dogmatic:

> And did those feet in ancient times trod on America's pastures
> of green?
> And did that anthropocentric God wane with his thoughts
> and beliefs so unseen?
> I don't think so. He's up there with the others laying low, vy-
> ing with those whom you've traded your life to to bless
> your soul.
> And have they told you how to think, cleansed your mind of
> sepsis and autonomy?

Or have you escaped scrutiny and regaled yourself with de-
pravity?
Now we all see religion's just synthetic frippery, unnecessary in
our expanding global cultural efficiency.
Now don't you fear this impasse you have built to your future?
Ever so near and oh so austere.

In my opinion, the worst line in this song is the one where I made a bold claim about religion ("religion's just synthetic frippery"). The rest of the song conjures up images that apply to everyone, regardless of whether they believe in God. And the most compelling lines of the song, in my opinion, are the questions. This song has been a perennial favorite among Bad Religion fans—believers and nonbelievers alike—and part of the reason for the song's success, I think, is that its questions are ones that listeners ask themselves.

Many Bad Religion songs could serve as a rallying cry for atheists, but Brett and I have always tried to shy away from blunt statements of revolution or cheap one-liners that lack depth or meaning. Painting a musical image with broad strokes allows listeners to arrive at their own conclusions. In the song "Atheist Peace" from the album *The Empire Strikes First*, I wrote:

Maybe it's too late for an intellectual debate, but a residue of
confusion remains.
Changing with the times and developmentally tortured minds
are the average citizen's sources of pain.
Tell me what we're fighting for, I don't remember anymore.
Only temporary reprieve

*And the world might cease if we fail to tame the beast. From
the faith that you release comes an atheist peace.*

*Political forces rent bitter cold winds of discontent and the
modern age emerged triumphantly.*

*But now it seems we've stalled and it's time to de-evolve and
relive the dark chapters of history.*

*Tell me what we're fighting for? No progress ever came from
war, only a false sense of increase.*

*And the world won't wait for the truth on a plate, but we're
ready now to feast on an atheist peace.*

This song could have veered into a scathing attack on all believers. But I wanted to try to motivate people rather than criticize them. The Socratic method of questioning always delivers better results than blunt attacks—at least it has for me. When people are encouraged to see the world as it is and wonder where it came from and what it means, they are ready to learn. Just as those who eventually turn away from religion begin by asking questions, those who raise questions can establish their worldviews on the basis of evidence and reason rather than dogma. They can engage the world directly without having to cut through a shroud of preconception.

I try to be careful in my assertions about the naturalist worldview. I realize that some people fear God, which biases their ability to be introspective or skeptical. Other people loathe God, which can impose a different set of obstacles to meaningful inquiry and dialogue. The God-fearers say that life on earth is guided by an intelligent being and that our ultimate focus should be on what happens to us when we die. The God-haters believe that religious believers are a bunch of brainwashed drones. Some of these God-haters have sought to form social groups of their own with guiding documents, gathering places (even if just on the Web), and a quasi-religious sense of community. The tacit ring of "come join us"[14] is apparent on their Web pages and in their public lectures. To me, these organizations of nonbelief can start to sound a lot like the ones they vilify.

Religions deal with ultimate questions: Where did I come from? What is my purpose in life? And what is the point in trying? There are so many of these "big picture" questions, and they are so difficult and timeless, that it is reasonable to assume that our most significant social institutions have developed to provide all-encompassing answers that satisfy most of them. The answers provided by religion, for instance, tend to be along the lines of "You are flawed, but don't worry because God made you that way," or "God wants you to live a good life," or "God is perfect and neither his will nor intelligence should ever be questioned." In other words, there isn't much room for debate when it comes to these ultimate questions.

Some of the solutions from the popular atheist literature can seem equally dogmatic. If you believe in God, you might be the victim of "moral terrorism,"[15] or possibly you are just brainwashed, but it's understandable, because your brain was "hardwired" that way when you were young and there's very little hope that it can ever be "rewired" when you are an adult. Reasoning such as this detracts from, rather than stimulates, the desire to keep exploring the big questions of life.

Yet these are exactly the questions we need to debate most fiercely. What does the fossil evidence say about where we came from? How far back can we trace the molecular origins of life? How can aging be slowed? What kind of life-forms might exist on other planets? All of the answers to these questions have "big picture" implications. And I believe there's just as much mystery surrounding these naturalistic realms of inquiry as there is in meditating or praying. But the process is different. The naturalist tradition is one of sharing information and comparing data. People from every corner of the planet can communicate with one another in the language of science. The polarity of the atheist/theist debate prevents this harmonious social activity.

It is time to cast aside the interminable and endless debate about God's existence. We need to focus on life and on the creativity—and inevitable tragedy—inherent in life.

CHAPTER 5

TRAGEDY: THE CONSTRUCTION OF A WORLDVIEW

[It] destroyed with one blow all the remains of my earlier dualistic worldview.
——Ernst Haeckel on the death of his wife in 1864[1]

We have lost the joy of the Household, and the solace of our old age.
——Charles Darwin on the death of his ten-year-old daughter, Annie, in 1851[2]

Think of the tragic events that have happened to you. Maybe it was the loss of a family member or friend. Maybe it was a missed opportunity due to some unforeseen circumstance. Perhaps it was getting fired from a job or a moment so embarrassing that it still causes you anguish. Tragedies large and small accumulate throughout a lifetime, and to stay sane we all have to make sense of them. Some scholars believe that tragic events in the personal lives of early evolutionists might have strongly influenced the development of their naturalist worldviews.[3]

Tragedy certainly has been one of the drivers of my intellectual development. My life has not been unusually tragic. I have been spared the traumas many of my friends have endured. But acknowledging tragedy has motivated me to both support others and reach out to others for support, and in so doing I have shared my difficulties and have listened to other people's. Tragedy is so common in our lives that I have come to see it as a familiar thread running through all living things. All organisms are affected by constant traumatic changes, which, in human terms, we interpret as tragedies. And a tragic sense of life is an inevitable accompaniment to a naturalist worldview.

+

On October 29, 1959, a car driven by Edward Michael Zerr, a prominent elder in the restorationist denomination known as the Churches of Christ, collided with another car in the small town of Martinsville, Indiana. Zerr suffered a broken arm, a broken leg, and several broken ribs, and on his way to the hospital he slipped into a coma. Four months later, at the age of eighty-two, he died without ever regaining consciousness.

Brother Zerr was a prolific writer, teacher, and preacher. He wrote articles for religious periodicals, biblical analyses of historical quotations, and two volumes of Bible questions for self-study. Over the course of sixty years, he preached more than eight thousand sermons from New England to California. He also wrote the music and lyrics for several religious songs, including two, "The True Riches" and "I Come to Thee," that are still sung in Churches of Christ services.

But the work for which E. M. Zerr is best known is his six-volume commentary on the Bible,[4] published between 1947 and

1955. He worked on the volumes six days a week, from four to eight A.M. each morning, while being supported by his church. The volumes are a marvel of textual analysis. It is clear that Zerr was intimately familiar with every word of the Old and New Testaments and could comment authoritatively on every nuance of biblical meaning. "It has been my constant purpose to avoid speculation," he wrote in the preface to the first volume, which covers the books Genesis to Ruth. "I have offered no explanation based on mere guesswork. In all instances where I was not sure I understood the passage, no comment was made. My sole aim has been to encourage a more complex understanding of the Sacred Volume. If such is accomplished I will consider myself fully repaid for all the time and labor expended."

E. M. Zerr was my great-grandfather and the father of my maternal grandma, who lived with us in Los Angeles during the winter months. My mother was never religious after I was born, but odd traces of Great-grandpa Zerr's beliefs still clung to our household. The Churches of Christ believe that music should only be sung with voices, not accompanied by instruments. The injunction against instruments came from a verse in Ephesians 5:19: "Addressing one another in psalms and hymns and spiritual songs, singing and making melody to the Lord with all your heart." These words are taken literally by my great-grandfather's denomination to mean that no instruments should accompany the purity of a human voice.[5] In fact, everything in the Bible is taken literally by his denomination. The Churches of Christ were part of the Restoration Movement, which grew out of the Second Great Awakening early in the nineteenth century. The movement sought a shift back to literal interpretations of the Bible and the restoration of the church established by Christ.

E. M. Zerr expected his grandchildren to follow the rules of his denomination while accompanying their parents to church, but my mom and her brother, Stanley, were allowed to play piano and guitar while they were growing up. Neither of them emerged from their childhood with much of a religious sensibility. But they both had and still have an abiding love of music.

The other family tradition that emerged from E. M. Zerr's devotion to biblical interpretation was the value accorded to a good education. Both my mom and Uncle Stanley went on to get PhDs, and their educational values were passed along to me, my brother, and my cousins. Academic discussions, along with musical sing-alongs, were among the highlights of our extended family gatherings at Mom's house for Thanksgivings. The discussions would start at breakfast and often carry on well into the afternoon. They were led by my mom and Uncle Stanley, but invited friends, opinionated teenagers, and even the kids were never discouraged from jumping into the fray. Many a tranquil autumn morning turned into an afternoon of heated debate about social policies or current events. But, invariably, the early evenings saw peace restored when Uncle Stanley broke out his guitar and banjo. Singing harmonies to those old tunes brought the family together.[6] I could never keep a sour attitude when the music began. It's impossible to be anything but happy when you hear the jangle of an old-timey banjo tune. All the cares of the world seem irrelevant.

Great-grandpa Zerr's death came too quickly and too abruptly for my family. I'm sure he would have seen me as a poor, lost sinner, but I know he would have loved participating in those sing-alongs. He was still healthy and active at the time of his accident. His death has always been a reminder for my family that tragedy can strike at any moment.

✦

Anyone who grew up in the L.A. punk scene of the 1980s knows that violence can be swift, unexpected, and tragic. My friends got beaten up regularly by the police during riots at concerts. Far more tragic were my friends' suicides or drug overdoses. I never used marijuana or any other drugs. My shock at being exposed to the drug culture when I moved to California made me vow never to get wasted like the kids in my middle school and high school. But my friends included me in their drug-abuse antics at punk hangouts on weekends. I felt uneasy being the sober watchdog but really wanted to be a part of the gang, so I didn't object at first. I went so far as to occasionally administer an injection of heroin or speed to my friends. It sends chills down my spine these days to think about what might have happened. I had no idea about doses. I could easily have administered a lethal dose to one of them and spent years in prison for killing a friend with drugs.

When I was in my early twenties, I heard from my bandmates: "Did you hear? Bob died last night of a heroin overdose." Then, just a few months later, "Tom killed himself yesterday when he crashed his car into the back of a parked tractor-trailer." I knew that Bob and Tom wouldn't have died if they weren't using drugs. Both of them were lighthearted and funny when they were straight. But when they used, they were secretive and reclusive. Their deaths made me reassess what I'd been doing. If drugs could change my friends' personalities so drastically, I wanted nothing to do with that scene. I didn't want to do those kinds of experiments with my life. The deaths of Bob and Tom, and the fact that I didn't have any genetic predisposition or psychological compulsion to use drugs, made me realize how lucky I'd been.

I define "tragedy" more broadly than do many people. I see it as an abrupt change in things we take for granted. It destroys lives and relationships that were supposed to be unbreakable. Tragic events are surprising, unforeseen upheavals that defy our understanding. There are no winners in a tragedy. There is only loss, stark and unforgiving.

Social relations are the source of many personal tragedies. The universal laws that bind people to one another emotionally serve to reinforce our beliefs about our interpersonal relationships. We rarely examine the unspoken rules of love, for instance.[7] But when love dies, or a marriage ends, or a child is estranged from a parent, we feel that the laws have been broken, and we sense a deep tragedy.

I have had my share of relationships that have ended in tragedy, which I attribute in part to my own lack of understanding. As a teenager, I assumed that the girl punkers in our crowd were the equals of the boys in every way. We were united by our desire to overcome societal pressures. The emphasis was always on the collective turmoil of our group and not on individual personal issues. (There were a lot more songs about "us," "our," and "we" than about "I," "mine," and "me.") My punk friends and I were better at practicing tolerance than understanding. Unfortunately, this also became an excuse not to address the different needs and perspectives of men and women. To this day, I am good at being tolerant of women's needs without fully understanding those needs. This has interfered with my ability to be a good husband. I have had a tendency to place my music, my education, and my research above the needs of my family. The breakup of my first marriage was a source of tremendous heartbreak for me. It made me realize for the first time the pain my parents had gone through when they divorced. It also made me vow to do better when the opportunity for love and marriage arose again.

Many people spend an inordinate amount of emotional effort on deluding themselves into thinking that they have not experienced any tragedy. They are afraid to acknowledge the sadness they feel because it can be all-consuming and interfere with normal social relations. In public, we are expected to swallow our sadness. We try not to cry or to express our sentiments too openly. This is part of a grand conspiracy. If everyone acknowledged the reality of tragedy, we would realize how painful life really is.

I have never felt comfortable deluding myself into thinking that my pain isn't so bad. If I feel a great loss, it gives me no comfort knowing that others have lost more, are more confused, or have had worse things happen to them. My feelings are my own, and I cannot compare them with other people's feelings. My pain and yours will always be different. We don't have any way of gauging its abundance. But ignoring it is nothing more than delusional.

Some people believe that too much public sentimentality interferes with others who are trying to overcome their own sadness. It's as if we are expected to help others deal with their fragile emotions while simultaneously ignoring our own desire for empathy. It is another way of denying the immediacy of tragedy. We have to acknowledge that there is no way of classifying pain except as a ubiquitous part of life that must be dealt with on a continual basis.

Tragedy is also an inherent part of nature and of evolution. No one can be a naturalist for long without recognizing the tremendous amount of waste, fear, and death that occurs in the biological world. Can a greater understanding of tragedy in the natural world help us deal with the tragedies that occur in our own lives? I have often asked myself this question. The study of fossils certainly has little to do with avoiding future tragedies. On the other hand, the study of history—such as gathering data on past civilizations and their interactions with the environment—can be very important when solving problems related to current ecological challenges.[8] If we agree that the ecological mistakes of our forebears can be avoided, then we have to conclude that there is a practical component to historical knowledge. Can this practicality be extended deeply into the realm of natural history? I have not arrived at an entirely satisfactory answer. But I have tried to answer it in the affirmative. I have studied the history of life for more than purely academic reasons. For me, the study of life's past has an emotional component as well.

✦

When Charles Darwin and Alfred Russel Wallace were crafting their theories of evolution, they were both influenced by an essay written by one of their countrymen. In 1798, the reverend Thomas Malthus wrote that the human population grew much faster than did the resources needed to feed it. The inevitable result for the human population was death and starvation. Malthus was a theist and believed that starvation, suffering, competition, and all the other evils in the world were placed here by God to teach us a lesson. Without the relentless threat of starvation and poverty, humans would have no reason to work hard and do well for themselves and others. He argued that widespread suffering for the good of the few was somehow the source of human virtue.

When Darwin and Wallace read this essay, they saw something different from Malthus. They independently realized that a mismatch between population and resources in nature could provide the mechanism behind what Darwin called natural selection. More organisms are born in nature than can survive and have offspring. As a result, organisms with the greatest fitness perpetuate their traits in future generations while unfit organisms starve, die of disease, or are eaten by others. Slight variations in the ability to procure needed resources lead some members of a population to survive and others to perish.

Seen in this light, the theory of evolution becomes a narrative justification for the presence of so much tragedy in nature. That seems to be how Darwin viewed evolution. The "struggle for existence," "survival of the fittest," and "competition for resources" all point to a perhaps unconscious effort to explain at least part of the suffering in the world. This view of evolution offers a justification for tragedy by exaggerating the role of fitness. It sees death as a by-product of one of evolution's main "mechanisms," which is the selection of only the fittest individuals in a population of organisms with varying traits.

But as we've seen, most of the variation in life is below the threshold of selection. Unfit traits and behaviors are a common part of life, including our own. Tragedy does not serve the higher purpose of clearing the way for individuals with greater fitness. It is far more random, senseless, and anarchic than most biologists presume.

A different observation does a better job of accounting for the presence of death. The world is a finite place. Over very long time periods, biological creativity must be balanced by biological destruction.[9] If this balance were not maintained, the world would choke on its own biological exuberance, or all living things would quickly disappear. At various times in the past, the forces of destruction have predominated. Yet after these periods the balance was quickly reestablished. From an evolutionary standpoint, tragedy requires and receives no justification. Death is simply the necessary complement to life.

The fossil record stands as inerrant testimony to the inescapable presence of death. The entire history of life on earth is chronicled by moments of death frozen in time as fossils. Each discovered fossil represents a unique remnant in a continuum of biological events, with death providing a snapshot of the past to its eventual discoverers.

Unexpected outcomes with mortal consequences typify biological analysis at every level—from the cell to the individual to the population to the species to the ecological community. At the lowest level of biological organization, single-celled organisms that split in two to reproduce—which were the only inhabitants of the earth for more than two billion years—could be said not to die. There is an unbroken lineage of these organisms from the origin of life to the present day. Single-celled organisms exist almost everywhere on our planet, from the dark recesses of our guts to deep subterranean rocks that never see the light of day to the ionosphere high above the earth. They were the first life-forms to appear on the planet and they have been a constant presence ever since. Single-celled organisms can die—when the environment changes, for example, or when they are eaten by other organisms. But I think of them more as tiny metabolic engines, always buzzing about, sometimes causing infection, more often doing things out of sight and out of mind that are crucially important for the sustenance of more recognizable forms of life.

Death became the familiar event of seemingly wasteful pointlessness, despite its biological necessity, when multicellular life evolved. Instead of single cells or small colonies of cells performing simple functions on their own, multicellular life brought with it complex tissues and organs all functioning in concert to the apparent benefit of a highly complicated organism. One characteristic of multicellular organisms is a "division of labor" between the body (somatic) cells and the sex cells (gametes). Gametes carry genetic information from one generation to the next. Once these cells initiate the development of offspring, the rest of the organism is no longer biologically necessary after the offspring is born. The somatic cells simply age and perish, and the organism eventually dies.

The evolution of multicellular organisms over the past billion years has made possible the development of a nervous system. These networks of specialized cells provided organisms with many specialized capabilities. Organisms with nervous systems can sense conditions in the environment with much greater specificity and sensitivity than other organisms. They can send signals from one part of the body to another at great speeds. The parts of the nervous system also can communicate with one another, giving rise in its most advanced form to the phenomenon we call "thought."

In addition, nervous systems can sense and register pain when an animal is injured. Nonhuman animals have varying degrees of what we call "consciousness," which means that they probably experience pain differently than we do.[10] But pain serves to remind them to avoid danger in the future.

The evolution of complex nervous systems also led to highly social animals, such as humans, chimpanzees, whales, elephants, and even the social insects. These organisms have varying abilities to consider themselves in relation to others. For instance, when death strikes and these animals lose an offspring, close relative, or companion, they experience the loss in different ways. Gazelles and other hoofed mammals are constantly preyed upon by carnivores such as lions or wolves. When members of their family or other members of the herd are killed, they show very little change in behavior. In contrast, other mammals, such as chimpanzees or elephants, can be so overcome by the loss of someone close to them that they fail to function normally. It's almost as if life to them is no longer worth living. Animals with more highly developed nervous systems—and especially those with a larger cerebral cortex—seem to perceive tragedy more profoundly.

Humans, with our highly developed cerebral cortexes, seem to be the only animals capable of conceptualizing the inevitability of death (although, of course, we do not know exactly what other animals with highly developed cerebral cortexes are thinking). This ability confers upon us unique forms of suffering. We recognize that everything around us must pass away. Yet that recognition provides us no solace, and when tragedy strikes, our pain is undiminished.

Suffering is an inevitable consequence of evolution. Naturalists see tragedy as an outgrowth of natural processes that have been occurring in multicellular organisms throughout history: bacterial parasitism, infant mortality, infection, starvation, catastrophe, species extinction. Does all this suffering serve any purpose other than reminding us to try to avoid suffering in the future? Perhaps it's too much to ask of any worldview—whether based on naturalism or religion—that it provide an ultimate answer to the question of tragedy.

+

Every once in a while, we are forced to confront the stupefying immensity of the universe. On summer nights in Wisconsin, during my childhood and adolescence, my brother and I sometimes had to escape the stifling mugginess of Dad's un-air-conditioned house, which was about a half mile from Lake Michigan and far enough from Milwaukee's and Chicago's urban lights to reveal a pitch-black sky. Occasionally we would lie flat on the front lawn, borrow Dad's binoculars, and gaze up at the stars. We knew very little about what we were viewing, but we had watched enough Carl Sagan to know that the lights we were seeing might have taken thousands of years to reach us, even though light travels nearly six trillion miles in a year. And the more we lay there the more stars we saw, extending away from us into the infinite blackness. I remember asking my brother, "How long has space been around?" He replied, "It's always been there." That answer was good enough for me as an inquisitive preteen, even though I now have a more complete and scientific answer.

My experiences on Dad's front lawn made it easier for me, when I began to study evolutionary biology seriously, to come to terms with something that most people find hard to conceptualize: the incredible immensity of time. People tend to think in terms of days, months, or perhaps decades at most. This is natural, because those are the time frames of human events that have personal significance. But that's like thinking of cosmic distances in terms of the roads we travel every day. Time stretches behind us as endlessly as does the black emptiness of space beyond us.

Most of us can trace our families back a few generations. But if we take twenty-five years as the average length of a human generation, then eighty generations separate us from the time of Christ. That's eighty cycles of births and deaths, eighty passages of DNA from one generation to the next, eighty opportunities for the line leading from our ancestors to us to go extinct.

Furthermore, 80 generations is no time at all, from an evolutionary perspective. About 8,000 human generations separate us from the origins of the anatomically modern humans who lived in eastern Africa 200,000 years ago. And 8,000 generations is a tiny number compared to the 250,000 human generations that separate us from our common ancestor with chimpanzees, or the 2.5 million generations that separate us from the extinction of the dinosaurs, or the 140 million human generations that separate us from the origins of life on earth.

We don't even have a good way of comprehending such huge numbers. Maybe we can get a grasp on a thousand—counting to one thousand takes twelve minutes or so. But even one million is a very difficult number to comprehend, despite the fact that we read about millions in newspapers every day. Once, on a road trip from L.A. to Berkeley, to play in a show, Brett and I were driving in his car and wondering about how to pass the time. We began jokingly singing the rousing verse, *"A million bottles of beer on the wall, a million bottles of beer . . . if one of those bottles should happen to fall? [long pause] . . . nine hundred ninety-nine thousand nine hundred and ninety-nine bottles of beer on the wall! [verse two] Nine hundred ninety-nine thousand nine hundred and ninety-nine bottles of beer on the wall, nine hundred ninety-nine thousand nine hundred and ninety-nine bottles of beer . . . if one of those bottles should happen to fall? [long pause] . . . Nine hundred ninety-nine thousand nine hundred and ninety-EIGHT bottles of beer on the wall!"* After about five minutes of this, our girlfriends in the backseat couldn't take it anymore and asked us to please stop. "But we've only gotten to 999,992!" we protested. Their request, however, led us to ponder just how long it would take to count to a million. So I counted for five minutes while Brett kept time. Then I restarted counting, this time from high to low. By this time, our passengers were disgusted with our nerdy game and began talking between themselves. (Those girlfriends didn't stick around long after that road trip.)

We ended up calculating that if two people did nothing but count for eight hours a day as their full-time jobs (with no lunch breaks, but weekends off), they would have to spend roughly twenty weeks of their lives to reach one million. That's five weeks longer than a university semester!

Scientific research has given us the remarkable ability to estimate when particular events occurred in the history of the universe, the solar system, and the earth. Radioactive atoms decay into other types of atoms at precise rates. By knowing how much of an isotope was present at some earlier time and how much is present today, scientists have come up with pretty solid numbers for the ages of the moon, the earth, and the first fossil evidence of life.[11] These dates cannot be calculated with absolute certainty, so geologists might speak of dates such as "1.65 billion years ago, plus or minus 82.5 million years" and feel fairly confident that the actual dates fall somewhere in that range. For example, based on several independent lines of evidence, the age of the earth has been estimated at 4.54 billion years, plus or minus 45 million years. That's a pretty amazing estimate, given the enormous amount of time that has passed since it occurred.

The oldest evidence for life on earth comes from fossils that are 3.4 billion years old, plus or minus 100 million years.[12] An incalculable chain of organic events leads from those earliest fossils to each of us, so each of us belongs to this process. "Finite but incalculable" is how I describe sequences of events in evolution.

The past is full of alien creatures, weird atmospheric conditions, and remote landscapes, as foreign to us as cosmic space and geologic time are to our human comprehension. For example, if you traveled back in time and went for a swim in a nearshore reef of the early Cambrian period, roughly 540 million years ago, you would enjoy the water—temperatures were about the same as in the tropical reefs of the Florida Keys, the Great Barrier Reef of Australia, and Cancún, Mexico. But if you were wearing a snorkel and mask and looked at the organisms in the shallow reef community beneath you, what you would see would look starkly different from anything in today's oceans. Reefs today consist mostly of organisms belonging to the animal phylum Cnidaria. These animals are part of a huge group that includes the jelly fishes, sea anemones, and the tiny, colony-forming, cup-shaped organisms that make up the calcareous framework of the reef communities in modern oceans. In the early Cambrian, reefs were composed of organisms called archeocyathids, which are now extinct. These creatures were close to sponges in their anatomy and classification. They were as different from the jellyfish and sea anemones that make up reefs today as humans are from clams. A few cnidarians populated Cambrian reefs, but they are all now extinct varieties.

If you got back in your time machine and snorkeled above a middle Mesozoic reef roughly 84 million years ago, you would see yet another strange sight. Where are the archeocyathids? What kind of cnidarians are these? This time, the organisms making up the reef would bear no resemblance to either sponge-like creatures or modern corals. The reefs of the late Cretaceous were composed of bivalves called rudistids. These animals had conical shells, roughly eight inches tall; they congregated in colonies and were held in place vertically by lime-rich nearshore marine sediment. No trace of rudistids can be found in any of the reefs preceding the middle Mesozoic era. Prior to the Jurassic, roughly 200 million years ago, they hadn't evolved. No wonder you didn't recognize them from your previous experience in the early Cambrian.

Reefs today are composed primarily of species that belong to a group of cnidarians called scleractinian corals. Scleractinians made up a very small portion of the reefs in the late Cretaceous. But 65 million years ago, during the mass extinction that also eliminated the dinosaurs, the rudistids became totally extinct. Both on land and in the sea, the extinction of major groups of organisms set the stage for a new era in evolution.

Mass extinctions are mysterious occurrences that have happened repeatedly in the history of life. They have been known for over 150 years. In 1360, a geologist drew an illustration that depicted what he believed to be a graph of the total number of animal species through geologic time, using fossils in England as his guide.[13] He had no grasp of absolute time, because there were no radiometric "clocks" at that time (radioactivity hadn't even been discovered yet). Still, he was able to show very clearly that two events in the past were responsible for wiping out a significant portion of all the species he was examining.[14]

Today we recognize those two events as highly significant landmarks in earth history. The most recent happened 65 million years ago and is known as the K-T extinction (after the terms Cretaceous, often abbreviated with the letter K, and the Tertiary period). The K-T extinction wiped out the dinosaurs, providing an opportunity for mammals to diversify and take over ecological niches formerly occupied by dinosaurs. The K-T extinction is now famous for the hypothesis that was advanced to explain it—a meteorite that crashed into the Yucatán Peninsula and plunged the world into catastrophic darkness.[15]

The Permo-Triassic extinction 252 million years ago (called "the Great Dying") was even more severe, eradicating 95 percent of all marine species on earth.[16] The Permo-Triassic extinction is thought to have been caused by massive chemical outpourings of greenhouse gases from volcanic activity in regions that would become Siberia. It took tens of millions of years for marine and terrestrial life to regain the levels of biodiversity present before the event.

Mass extinctions cause death across a wide spectrum of living things. Even very successful, well-adapted species and individuals succumb. It is generally impossible to predict which species will be eliminated in a mass extinction.[17] Fitness plays little or no role in these extinctions. When tragedy strikes, no individual is prepared. It appears to be a simple matter of luck which individuals survive.

After a mass extinction, species again diversify. But these species can look quite different from what came before. The most obvious example is the rise of mammals after the extinction of the dinosaurs. As with extinction, it can be very difficult to predict which evolutionary lines will flourish after a mass extinction. Life often takes on a different complexion after tragedy.

Could all these ecological replacements have some grand cosmic significance? The more I've studied paleontology, the more it seems crystal clear to me that the answer is no. Life isn't guided by any purposive forces. This or that particular phylum might dominate during one phase of history only to be completely wiped out and replaced by organisms from a different phylum. Even human beings, from this perspective, appear to be an unpredictable product of evolution, not the evolutionary pinnacle toward which a steady succession of inferior life-forms was aimed.[18]

There is an immense history of life that needs to be explained if God is the Creator. Before creating humans, God would have done a tremendous amount of seemingly pointless experiments with living creatures, causing mass extinctions and limitless pain and suffering. How caring and wise was that? It's hard to be a theist after spending much time with the fossil record.

✦

Anyone who is a theist and believes in a caring, responsive, and powerful God must come to terms with religion's central problem: the presence of so much suffering and misery in nature. The paradox can be stated succinctly: if God made us in his own image, which implies that God must feel compassion for other entities, then how can he allow so much tragedy to occur in the world?

All major religions have sought to answer this question. Humans are inherently sinful and evil, they say, or evil is a necessary prerequisite for human free will. In some religions, evil is a form of punishment from God, while other schools of thought hold that the ways of God are simply unknowable. Human beings have had thousands of years to reconcile their experience of the world with their religious beliefs. Yet none of the explanations they have devised is either comforting or satisfactory.

Many people use the precepts of their religion to explain tragedy. They say that a person's death came because "God called" or that a person is in a better place after he or she dies. I dislike these explanations. They appeal to shallow parts of our empathy and direct our gaze away from the inevitability of tragedy and loss. Sometimes, at a funeral, a person will say to me, "He's in heaven with his loved ones now," or "Well, at least he lived a good life." I don't know what to say in response to such statements. I don't believe people go to heaven, and maybe the person saying it doesn't believe that either. Such statements remind me that everyone is struggling to make sense of the inherent tragedy of life.

Such statements also remind me that we have to be very careful in applying the principle that only God knows when it's someone's "time to go." If taken too far, such statements can offer a justification for taking irresponsible actions. I'm sure that many of my friends who abused drugs did so by maintaining careless philosophies: "Hey, I know I'm abusing, but I don't control when my time is up!" In human life, many deaths are preventable, such as many of the senseless deaths caused by car crashes, or thousands of the deaths caused by firearms in the United States, or part of the toll from infectious diseases. Death may be ultimately unpredictable, but the odds are stacked against us if we fail to take reasonable steps like fastening seat belts, locking guns in cabinets, and using condoms. Failure to do so represents the sordid flip side of religious explanations that seek to justify evil. They can seem to imply that all tragic events are part of God's plan and that we should not bother doing our best to minimize their occurrence.

In Western religions, the central metaphor for tragedy is sin. Humans are different from their Creator because God is perfect and we are not. God, however, gave us "free will" because we are his favorite creation. Humans must exert their free will to do good, despite being inherently sinful, in order to gain entry into heaven and live an eternal afterlife in paradise. Sin then becomes the justification for blame and punishment. You deserve to be punished if you don't exercise your free will to overcome your inherently sinful nature.[19]

When tragedies befall others, we tend to look for the ways in which these others sinned instead of recognizing and considering the long chain of events that led up to the event in question. Was a robber desperate to feed a family? Was a murderer beset by insane jealousy or greed? Was a man's violent behavior predictable from his surroundings or experiences? If such influences were taken into account, perhaps a more meaningful program of rehabilitation could be created instead of the strictly punitive institutions we have today, where prisoners grow old while pacing off a useless life until they die in a cement cell. Don't get me wrong, I believe that many criminals are not fit for society and need to be locked away. But that's all the more reason why we need excellent rehabilitation facilities to reeducate those who are fit for society and keep the others forever sequestered where they can do no harm.

Our penal system reflects our need to construct stories that explain tragedy, even when tragedy is ultimately pointless. These "central narratives" form the comforting and profoundly meaningful basis of our worldviews. One of the most common central narratives of modern-day America involves sin and its relation to blame. Our fascination with blame feeds our passion for punishment. People are told as children that punishment is the consequence of choosing to do the wrong thing—an outgrowth of our unsubstantiated belief in free will. Because of the power of this message, we obsess over punishment. Our news stories, movies, and television shows all subscribe to the same central narrative and feature people who outrage society by their hideous acts of sin. Pick a story at random from your favorite newspaper, movie, or book, and punishment and sin are likely to be threaded throughout its fabric.

There is, however, a double standard in our central narratives. We only feel compelled to address the tragic side of life. When things are going great, we don't try to make sense of it. But when tragedy strikes, we beat ourselves up and tend to default to our worldview for answers. It seems natural to want to know how the pain was caused and what events led up to it.

Can an understanding of tragedy's ubiquity ease our pain? No. Pain cannot be analyzed away. There is no reply to tragedy. Learning about the causes of a tragedy does not erase the sadness. I don't know all the causes that led to my friends' tragic deaths. I only know that dying is a part of life and that I am glad that we were able to spend part of our lives together.

I try to recognize tragic events as typical of life, not as anomalies and not as purely depressing misfortune. And I do try to understand what happened in a tragedy and how it is different from what I expected. Such understanding is at least partial compensation for self-awareness. Understanding also gives me empathy for other people's tragedies, because there is no better way to gain empathy for the pain of others than to acknowledge life's painful experiences as they affect you.

The naturalist perspective offers only an analogy to the tragic events of our own lives. Yet I try to remember that tragedy has occurred since life began. Sometimes I think about what mass extinctions reveal about life. Life doesn't end after such events, but it takes on a new complexion. After a tragedy, my community and family roles might change. But I should not obsess over any single traumatic experience. I must remember that tragedies occupy just a moment in time, and the here and now is the product of an incalculable accumulation of tragic events.

Life is best seen as a series of tragedies marked by fitful progress and recurring setbacks. There is as much disappointment as joy. But tragedies need not cause despair. They can remind us about the realities of the natural world of which we are all a part.

CHAPTER 6

CREATIVITY, NOT CREATION

[F]acts of nature frequently fail to accord either with the wishes or with the apparently logical preconceptions of human beings.

—Julian Huxley[1]

Natural selection . . . has no purpose in mind. It has no mind and no mind's eye. It does not plan for the future. It has no vision, no foresight, no sight at all. If it can be said to play the role of watchmaker in nature, it is the *blind* watchmaker.

—Richard Dawkins[2]

In 1980, when I was fifteen, Bad Religion had been practicing for a few months to perfect six songs that we were going to release as an EP. When we arrived at the recording studio, the producer asked us, "Are you guys a power trio?" The four of us looked at one another with puzzled amusement to make sure that all four of us were, in fact, present, but then we answered yes. I guess we thought that "power trio" meant something in recording parlance that we were too young to understand, and we didn't want to look like we didn't know what we were doing.

When we played our first song, it immediately became apparent that the producer had no idea what we were trying to accomplish. He had never heard punk rock before, and he insisted that our songs weren't finished. "You need to put a guitar solo in the middle. And the song needs a chorus. Then it will be done." But we said that there was no guitar solo in this song and that the song was done. The producer just rolled his eyes and continued on with the session, even though I'm sure he thought we were a bunch of teenage punk know-nothings.

Those songs are still in print and have sold hundreds of thousands of copies. They have never been criticized for lacking guitar solos or choruses. We didn't have any way of gauging the potential of our creative efforts. We were novices in the practice of songwriting and recording. But we had chanced upon a fortunate combination of style and lyrical hooks that would come to be judged as successful. Maybe we would have been just as successful if we had listened to the producer that day. But there was a disconnect in our belief systems. He thought we were trying to accomplish something that required strict adherence to a rigid formula. We believed that we were accomplishing something more vital by rejecting traditional rock-and-roll song elements. We were hell-bent on overturning standard practices, and in the process a unique sound emerged.

Creativity is often misperceived as something that has been designed or intended.[3] In fact, truly novel and lasting innovations are often surprises, and they startle us because they are surprising.

Some people have no desire to be creative. They believe that if they follow the rules and routines, they will be able to claim that they have lived a successful life. Maybe they think that, by doing so, they will have achieved some utilitarian goal or useful end. But I believe they will have achieved only a fleeting taste of success.

Lasting success requires creativity, even if most creative feats are ultimately accidental and unpredictable. Rules and routines may be tolerable or even comforting in the short term. But eventually they need to be scrutinized and in many cases rejected to make intellectual or emotional progress. Rebellion has to be part of the response to rigid social institutions, or stagnation is assured. If evolution teaches us anything, it's that life is in a state of constant change. There is anarchy in the variation that serves as one driver of evolution, and there is anarchy in the inability of life to remain static. Eventually, radical changes beset every living thing.

Institutions that enforce rigid adherence to their own tenets must be scrutinized with particular skepticism. Religions, political parties, corporations, and even bands can fall into the trap of demanding loyal and unwavering devotion. They can require that followers adopt not just a specific way of acting but a specific way of thinking. Institutions, by and large, strive for permanence, and they almost always see life through a formulaic lens and strongly disfavor individuality and change.

I have found this to be the case even in punk rock. Sometimes punkers will say to me that they used to be fans of Bad Religion until we let them down by releasing an album that didn't fit their definition of punk. We have released fifteen albums so far, which means that people have had plenty of opportunities to become indignant over our music. I think our recent albums are just as confrontational and challenging as our earliest ones. But our music has changed over time. We have become better craftsmen. We have broadened our emotional range. We have found new sources of inspiration and creative innovation.

Many people think that only artists are creative. But, in fact, all of our lives are creative works in progress. Each one of us has the potential for creativity. If you have children, the surprising, unexpected traits of your offspring represent an inevitable result of biological creativity. Or you may modify the environment around you, including the social environment. The creativity inherent in life is the counterbalance to tragedy. It affirms our belief that life is a good thing and provides a rich potential source of human meaning.

Religions find the source of all life in God the Creator. Individuals may differ in the extent to which they attribute individual acts of creation to God. They may believe that God created the universe and then let matter and energy run their course. Or they may think that God is responsible for the movements of every molecule. But believers are united in the conviction that without God there would be no creation, no past, no future.

Naturalists find creativity in the physical universe, not in the workings of an immaterial god. In the naturalist worldview, creativity emerges from natural laws that operate spontaneously. Matter and energy come together to produce an endless diversity of physical forms and phenomena, some familiar to us, some strange and unexpected.

The view that the physical world can be creative has always raised problems for many people. How can something new emerge from the blind collisions of purposeless atoms? In particular, how can life, including organisms capable of feeling love and fear and ambition, emerge from the inanimate detritus of a lifeless planet? Answering these questions—even in part—requires taking a very brief look at a very large topic: the history of the universe.

✦

Given the rate at which galaxies are rushing away from one another and the dim glow of radiation that permeates space, all of the matter and energy in the universe appears to have originated about 13.5 billion years ago, in an event known as the Big Bang. From an infinitely small and infinitely dense point, the constituents of the universe burst into existence and rapidly expanded, resulting in not just all of the matter and energy we see today but space and time as well.

To many people, the obvious absurdities of the Big Bang—such as the idea of an infinitely dense point of matter—reveal creation by the hand of God. As far as I'm concerned, they are free to think so. Naturalists have little to say about what preceded the Big Bang, largely because the event itself seems to have destroyed all evidence of what came before. But even the Big Bang is not exempt from scientific examination. Physicists are currently using the largest, most expensive scientific apparatus ever built, the Large Hadron Collider in Europe, to smash subatomic particles into one another with tremendous force to investigate what occurred during the first moments of the Big Bang. These experiments could lead to a much better idea of the moment that creativity began in the universe. They may show there has been an infinite cycle of universal explosions and collapses, or that multiple universes exist alongside one another, or that nothing existed before the Big Bang. Personally, I don't spend much time worrying about this question. I'm content to view the 13.5 billion years since the Big Bang as a period of immense creative ferment that will never be fully explored.

Once the matter created by the Big Bang cooled sufficiently, the universe consisted of a vast cloud of hydrogen atoms—consisting of a single proton surrounded by a single electron—along with a smattering of slightly heavier elements, including helium (with two protons) and lithium (with three). The universe at that time was about the most boring place imaginable. It consisted of nothing but disembodied atoms drifting through space and the radiation left over from the Big Bang. Yet the potential for infinite creativity was already present in the most unassuming of places: the asymmetries of the hydrogen atom. A hydrogen atom is not a featureless round ball that just bounces off other hydrogen atoms. It has an inherent duality, a positive proton and a negative electron, two polar opposites. Both the proton and the proton-electron system can acquire energy, which causes them to assume different configurations. All of the creativity we observe in the world arises ultimately from the potential shapes inherent in hydrogen atoms.

Early on, as now, matter in the universe was not uniformly distributed. Some parts of the universe had more atoms and some had less. The atoms in the denser regions began to attract one another gravitationally. Soon they began to clump together into spinning balls and disks. Inside these balls of gas, pressures and temperatures became so high that the hydrogen atoms began to collide with one another with tremendous force. (Actually, these "atoms" were charged ions, because the temperatures were too hot for electrons to stay attached to protons. But for the purposes of this description, I'm going to refer to both charged ions and neutral atoms as atoms.) When that happens, two hydrogen atoms (well, really ions, but you get the point) can fuse to create a helium atom. This process, called nuclear fusion, releases immense amount of energy—it's the source of energy in hydrogen bombs. Because of this release of energy, the compressing balls of gas began to shine, and the first stars blazed with light.[4]

Once the first generation of stars converted most of their hydrogen into helium, they began to contract again, and the internal heat and pressure again rose. Eventually, their helium atoms began to fuse with one another and with other atoms, creating heavier elements—oxygen, carbon, silicon, iron. The stars then entered their death throes. Some shed their outer layers into space. Others self-destructed in the massive explosions known as supernovas, which created elements even heavier than iron—copper, gold, lead, all the way up to uranium.

Our own solar system formed from a spinning disk of gas and dust that was enriched with these heavier elements. When our sun began to shine, its radiation heated up and dispersed most of the lighter elements. The heavier elements then clumped together to form the rocky inner planets of our solar system—Mercury, Venus, Earth, and Mars. By about 4.5 billion years ago, the newly switched-on sun was surrounded by a suite of stately, though lifeless, planets.

Within less than a billion years, according to chemical evidence found in earth's oldest rocks, the earth was occupied by living things—cyanobacteria, the 3.4-billion-year-old fossils mentioned in the chapter 5. Of course, the first appearance of a fossil doesn't necessarily coincide with the origin of the organism it represents. Life on earth didn't suddenly originate with fully formed bacterial filaments.[5] It probably went through numerous stages of evolution that were not preserved as fossils. Before bacteria, there might have been cell-like structures with membranes but no ability to self-replicate. The earliest DNA might have been free-floating and not protected by any membranes, which would not have left fossils. The earliest appearance in the fossil record has to be taken as a minimum possible age for an organism, and the actual age is unknown.

When religious believers cite a possible place for God's intervention in the universe, many point to the origins of life. It seems like a clear dividing line in earth's history. Before that, the earth was a barren ball of rock spinning through space. Afterward, it contained organisms, entities that were fundamentally different from anything that had existed previously.

The origin of life is one of the toughest problems in science. It took place billions of years ago, and evidence bearing on the problem is vanishingly scarce. Yet the overall outlines of life on the early earth—even if not the exact details—are gradually yielding to scientific investigation.[6]

Because hydrogen atoms and other kinds of atoms have complementary shapes, they can stick together when they collide. The result is a molecule—a collection of two or more atoms held together by their shape-dependent attractions. Given their shapes, some atoms are more likely to stick together than others. And they are especially likely to do so when the presence of another molecule, known as a catalyst, increases the likelihood of preexisting atoms or molecules colliding in the proper configuration.

Some catalysts can create more of themselves—a process known as autocatalysis.[7] In such cases, they can use freely dispersed atoms and molecules to form concentrations of particular kinds of autocatalysts, some of which have the potential to form microscopic structures shaped like balls, tubes, or polyhedrons. Such molecules are said to be self-assembling. An example is the shell of a virus, which is made up of proteins that self-assemble into the hollow structure that encases a virus's genetic material.

Imagine a collection of various autocatalytic molecules float-

ing in a warm soup somewhere on the early earth. If some of
the molecules self-assembled into hollow structures, they could
enclose a sample of the autocatalytic molecules in the soup. This
"proto-cell" could contain everything it needs to make more of
itself. It would need a constant supply of freely disbursed atoms
and molecules, but it could get these by breaking open and re-
forming or perhaps through some sort of opening in its protec-
tive enclosure. A proto-cell like this could repair itself if it broke
apart. It also could replicate itself. These are the theoretical stages
that could have led to the primitive life-forms preserved as earth's
earliest fossils.

There is no reason to assume that only one set of molecules
could engage in this process of autocatalysis and self-assembly.
There may have been numerous creative combinations that didn't
evolve. Significantly, however, different sets of molecules could
have engaged in similar processes. At this point, all of the basic
ingredients for Darwinian evolution were present. The proto-
cells that were most successful in acquiring resources from the
environment and reproducing themselves would have become
more numerous. Changing molecular combinations could have
resulted in a primitive form of variation, another key factor in
evolution.

It's a long way from proto-cells to human beings—or even
from proto-cells to the earliest forms of life seen in the fossil
record. But it marks the beginning of the evolutionary process.
Proto-cells have many of the features of life, even if their claim to
such a distinction is dubious. They thrive in good environments
and go extinct when their environments experience catastrophic
change. These exhibit the dimly lit, theoretical origins of what
today we recognize as creativity and tragedy. They are individual

entities—true "selves." They persist for a time and spontaneously give rise to descendants that can incorporate new chemical reactions and molecular structures.

There is no guarantee that life began this way. Yet it is an entirely plausible account of life's origins that does not require divine intervention to breathe life into the dust of the earth. In fact, nowhere in this brief history of the universe is God's intervention necessary. It can be futile for religious believers to point to specific historical events as proof of God's existence, given the propensity of science to develop nontheistic explanations for those events. All such discussions must begin and end with a statement of one's worldview, monist or dualist, theist or naturalist.

✦

From my earliest childhood, my parents encouraged and highly valued creativity. My dad's favorite gifts around holidays and birthdays were oil paints, sable brushes, and plastic models of aircraft or hot rods. More often than not, I painted surreal pictures that involved ideas borrowed from my favorite music albums. I also spent time with my friends thinking of the most creative ways to dismantle the plastic models after they had stood on the display shelf long enough. This usually entailed skillful blasting with a BB gun at ten paces in the makeshift shooting gallery we had in Dad's basement.

As very young kids, we were encouraged to use the dictionary to "try to figure it out for yourselves first." This inevitably led to some nerdy games. My best friend, Wryebo, and I were rarely able to stump each other when we challenged each other to find obscure words in dictionaries. We were also shocked at some of the bad words we could find, and we liked to impress our friends with the anatomical or occupational definitions of such common slang. One time I challenged Wryebo to look up the word for the female equivalent of "penis." He looked at me a bit confused. "Yes, I finally stumped you!" I exclaimed. "No you didn't," he said, "I'm just trying to think how to spell it." At that point, I decided it should be a team effort to find the definition. "Penis" was easy to find, but the word for female genitals was much more elusive. After about ten minutes without success, we both decided it was time to bring this puzzle to the attention of Wryebo's mother, an academic who was rarely nonplussed. Wryebo handed her the dictionary and said, "Mom, we found 'penis' but we can't find the word for the ladies' one. How come?" "Well, boys, maybe you weren't spelling it correctly . . . here it is." She read to us, "A canal leading from the vulva to the cervix of the uterus in women and most female mammals." Through our snickers and giggles, we still were puzzled. Wryebo asked, "But, Mom, how did you find the definition way near the back of the dictionary for a word that starts with B?" "What word did you boys think you were looking for?" We responded in unison: "Bajina!"

Wryebo and I spent most of our grade-school years inventing competitive games out of whatever artifacts or unused implements we could find around our houses. Hallway broom-handle hockey, aluminum-foilball, footskit (a hybrid of football and basketball), and flip-the-creamer were just a few of the games we played. This last game was played with the one-ounce plastic cream containers our parents kept in the fridge and served with coffee. (My family still plays the game at diners when we order coffee and the cream is served in those one-ounce cups. The conical shape of the cream containers makes them relatively stable when turned upside-down on the table. The game consists of seeing how many times you can flip the creamer 360 degrees in the air from its resting position to the same position. The most consecutive successful flips wins.) We rarely found a shortage of raw materials to stimulate our competitive imagination. Sometimes, if we ran out of model airplanes or hot rods to shoot, we would simply play blast-the-model-paint-jars in my dad's basement shooting range.

I was always looking for ways to combine ideas, activities, and discarded or unused things, usually just to pass the time. I enjoyed making "log houses" out of sticks. When it rained and the street gutters filled with rushing water, I held "canoe races" using leaves of various shapes and sizes. When I painted in my room, I often used "mixed media" (paint, glue, fabric, magazines, and newspaper) to come up with posters that resembled the pop-art books that my dad kept in his library. I never believed that I was inventing or revolutionizing anything. But I believed that my creations were good, and the process of making them was a satisfying way to pass my afternoons.

Some of these childhood creative experiments became pastimes, activities to which my friends and I returned over and over. Others were merely ways to beat boredom on a particular day. Only a handful of our creative combinations became pastimes that stuck, like flip-the-creamer, which we have passed on to our own children. Footskit went extinct long ago.

✦

I did a lot of passing the time during my childhood. Maybe as a result, I wasn't much of a student in junior high school, and my grades in high school were, for the most part, lousy. Only after I began reading about evolution did I start working hard in my classes. My last semester at El Camino Real High School, I finally achieved what I failed to do in all previous semesters—I made all A's.

Because of my undistinguished high school record, I wasn't ready to attend a research university where I could immediately take hard-core science courses and do research. But thanks to some string-pulling by our drummer's father, who was a professor at California State University at Northridge, I was admitted there for the fall semester after I graduated from high school. I learned a lot from the excellent professors at CSUN. Though mostly known as a prestigious "state school" for teachers, it also had a top-notch faculty of geologists and biologists who inspired me to go outside and study the great natural surroundings of California. That's when I began doing the mountaineering and wilderness exploration that quickly evolved into an insatiable craving for outdoor adventure. I worked hard at Northridge, mostly taking all of the science classes I had so assiduously ignored in high school. The following fall, after a brief stint at the University of Wisconsin, Madison (which I had to leave because I couldn't qualify for in-state tuition), I applied to transfer to UCLA and was accepted.

By that time—the fall of 1984—the punk scene was pretty much dead in Southern California. I loved the music and the performing. But I was traumatized by the violence surrounding punk and deeply disturbed by the constant association of punk with nihilism and hatred. The punk scene of 1980 was full of tolerant, earnest kids who were interested in current events and weren't afraid to challenge the norms in ways that were intellectually and artistically provocative. But the scene fell apart from its own popularity. More and more people were coming to shows who didn't appreciate punk's original values, like individuality, self-expression, and artistic creativity. Ironically, the newer and larger crowds wanted more conformity and predictability from the bands. What could only be described as hooliganism started to predominate at punk-rock venues. Some of the bands started to take on the ganglike mentality of their fans. The shows began to get violent, and promoters didn't want to allow punk bands into their venues.

Most of my friends who were in punk bands in the early 1980s gave up trying to make punk viable. But a lot of them still loved music, and they continued to play in bands. Usually this entailed growing and teasing their hair, putting on women's makeup, and singing with high, squealing voices. The "hair metal" of the 1980s began to flourish at the time of the demise of the punk scene. The same venues that would put on punk shows in the early 1980s switched their focus and had great success showcasing bands like Guns N' Roses, Faster Pussycat, Mötley Crüe, and Ratt. Many of the fans that began to take an interest in these bands had been punk rockers only a couple of years earlier. To them, punk rock hadn't so much died as it had transformed into Hollywood hair metal. So, if you were a young person in L.A. who was into underground music between the years 1983 and 1988, your song choices were either shouty anthems that promoted group mentality and strength in numbers (from ganglike bands of the tattered punk scene) or pouty, high-pitched melodious anthems and ballads that depicted maudlin montages of relationships and rock-and-roll lifestyles (from the hair metal bands).

I might have drifted away from punk music altogether, as many punkers did, if it hadn't been for Greg Hetson. Every once in a while, he would call me up and say, "Hey, if you want, we can get the horn section together and go play a show at club XYZ this weekend." Bad Religion had no horn section, of course, but the implication was that since there was no real punk scene anymore, we could just throw together some players and have a fun gig, playing our old songs from years past for a one-night stand. It turned out that a small, loyal following would still come to see Bad Religion as late as 1986, even though we hadn't put out a complete punk album for four years. We did record a six-song EP, *Back to the Known*, in 1985, which Brett produced but on which he didn't play guitar. The EP repudiated the experimentation of 1983's *Into the Unknown* by returning to punk-rock songs, and small independent record stores sold it in California, Arizona, and Nevada. (When Greg Hetson heard *Into the Unknown*, he knew punk rockers would hate it. His critique was a classic blend of consolation and negative judgment: "Fuck 'em if they can't take a joke.") Basically, Greg Hetson and I were keeping the band together mostly to satisfy our few loyal fans, who were still interested in thought-provoking songs about philosophy and current events. Except for very occasional weekend shows in 1985 and 1986, Bad Religion was essentially in hibernation.

I graduated with a bachelor's degree from UCLA in the winter of 1987 and immediately landed a job at the L.A. County Museum of Natural History, where I had volunteered in high school. My official job title was "assistant preparator" for a curator who collected fossils on expeditions in South America. As in high school, my job was to carefully separate the bones from their rock matrix, using a combination of dental tools, handheld grinders, micro–sand blasters, and an impregnating glue called Glyptol. It often took me weeks to prepare a single reptilian jaw or mammalian skull for description, but my pace was adequate for a beginner, and I considered this a challenging and creative exercise of my interests. As I spent hours with the fossil-prep tool kits, carefully uncovering bones from their matrix, I felt the same contentment that I experienced concocting "mixed media" paintings with the brushes and tools my dad gave me for birthday presents. But I could only take a few hours at a time of sitting still. My restlessness constantly asserted itself.

When I graduated, I was already committed to beginning a master's program at UCLA in the fall, but I had mixed emotions about my future. I knew that jobs in natural history were very rare and that I was lucky to have one. But despite my feeling of satisfaction, I knew that a laboratory job would never fulfill my desire to explore. As an undergraduate, I had scoured the university catalog for classes whose description included the words "fieldwork required." I wanted to get off campus as much as possible and study nature firsthand. In many of my courses, we would hear morning lectures about general principles of natural science and then jump into departmental vehicles in the afternoon to see examples of what we'd just learned while the concepts were still fresh in our memory. And many of my classes had at least one extended excursion that required lots of hiking and camping.

Fieldwork often appears to be well thought-out and purposeful, but most of my "aha!" moments have come simply from being open to whatever observations come my way. For instance, I stumbled onto the *Atta* ant project by hiking around in the forest and looking at leaf litter. Studying ants more deeply led me to the idea that natural selection might not be as powerful as I had thought, which, in turn, opened up an entirely new literature and way of thinking for me. When one observes nature with an open mind, there is limitless opportunity for learning and for creative intellectual reflection.

✦

As I settled into my job at the museum, I began to romanticize the great experiences of naturalists. I spent long hours studying the skulls and skins brought back from past expeditions, reading field notes of dead explorers, and scrutinizing their locality maps. In the countless drawers of the museum collection, I pored over specimen labels that read "Malay Archipelago, 1928," "Marshall Islands, 1957," "Cook Inlet, 1940," "Yukon Delta, 1976," "Brahmaputra Drainage, 1955." The geographic names and specimens that went with them evoked in me a great longing to explore. In those museum drawers I saw a reason to visit remote places and come back with things that might contribute to human understanding of the natural world.

One day, while discussing my future research plans, the curator who was my supervisor mentioned that he was looking for an assistant to join him on his next expedition. He was planning to collect fossils in the Amazon Basin, but he had also secured funds from the museum's Division of Birds and Mammals to bring along an extra person. I didn't think about it for a second before saying I wanted to go. I told him that I had taken numerous field-oriented classes as an undergraduate, in addition to mammalogy, ornithology, and ichthyology (the study of mammals, birds, and fishes, respectively). I sold myself as perfectly suited to do whatever was required. Before the end of the meeting, he had essentially hired me for a twelve-week Amazonian adventure.

I had six weeks to get ready for the trip. During that time, my instructions were clear: spend every day in the Division of Birds and Mammals and learn how to prepare wild specimens for museum curation. One day, I was heading out the door of the division when the curator asked, "Oh, and by the way, you do know how to use a shotgun, right?"

"No problem" was my response, even though the most I had ever done before was shoot model airplanes and enamel paint jars in my dad's basement. But I wanted to be a member of that expedition far more than I was worried about my lack of experience with a gun. I figured this was going to be my only chance to see the Amazon Basin.

Unfortunately, as I spent more time in the Division of Birds and Mammals, I came to a disturbing realization. One of the goals of the L.A. County Museum of Natural History was to help establish a vast region of the Amazon Basin in northern Bolivia as a protected conservation zone. In order to protect a habitat, you have to make the case that it contains valuable flora and fauna. That meant collecting specimens and bringing their carcasses back as evidence. By sacrificing some animals, you could help save the species at large. My actual title on the expedition was "collector of birds and mammals." That meant I had to shoot, trap, snare, and kill almost everything that moved.

My training went well. Soon I could skin a small mammal, flense its carcass, and fill out the species identification tag—all in about fifteen minutes. Birds took me a bit longer, but the results were the same: beautifully prepared skins and complete skeletons. My notebooks were full of instructions and templates for recording habitat and species data. I was intellectually prepared to be the best field biologist that the curator ever brought along on any tropical expedition. And I felt that I would soon be contributing to a much larger cause: the documentation and preservation of biodiversity. I had no inkling of the experiences that were waiting for me.

✚

Expeditions sound romantic to almost everyone. As we checked in for our flight at Los Angeles International Airport, the ticket agent asked us about the special wooden crates that we had checked as baggage. The curator said, "We're going to dig dinosaurs in Bolivia. It's specialized scientific equipment."

"Oh my goodness, dinosaurs!" replied the ticket agent. "Kim, these men are scientists who dig up dinosaurs, isn't that exciting?" I felt vaguely uneasy about the lie told by the curator, but I didn't want to ruin Kim's day by telling her the truth: "I'm going to the jungle to blow the brains out of anything that moves for the next twelve weeks!" Actually, I still didn't know much about our itinerary. I knew we were going to fly through Miami to La Paz, the capital of Bolivia, where the curator was going to acquire the collecting permits and permissions needed to explore remote regions of the country. That would take about five days, so I knew I had five days of "city life" before we headed into the jungle.

Landing in La Paz was otherworldly to me. It was my first visit to a place where I felt like an outsider. I have since had that feeling many times from traveling to foreign countries with Bad Religion, but Bolivia was my first. I wasn't prepared for the poverty that surrounded our downtown hotel, or the drab and uncomfortable accommodations, or the cold and polluted air, or the freeze-dried potatoes served at every meal. But I especially wasn't prepared for something I should have thought about: the altitude.

La Paz is located at an elevation of about twelve thousand feet in the Andes Mountains. It is the highest capital city in the world. When you are flying there from sea level, this drastic change in altitude can be dangerous. One thing all mountain climbers know is that there is less oxygen in the air at higher altitudes, so for your muscles and brain to function properly, you have to breathe faster to absorb oxygen. If your body doesn't adjust properly to the drastic decrease in oxygen, terrible side effects, collectively known as "altitude sickness," can set in. I had always taken necessary precautions when I did fieldwork in the Sierra Nevada Mountains of California. I would spend a day at around eight thousand feet and only slowly make my way up to eleven thousand or twelve thousand feet over the course of the next few days. But there was no time to get used to the change in altitude during the eleven-hour flight from Miami.

When I went to gather my luggage at the airport after we landed, a porter snatched away my suitcase and threw it into a waiting taxi. I appreciated this, because just the walk from the plane to baggage claim took my breath away. As soon as I entered the hotel, an attendant gave me a ceramic cup of hot water containing a few floating dried leaves. *"Aquí lo tiene, señor, coca te."* Coca? As in cocaine? As it turned out, one way Bolivians help foreigners adjust to high-altitude effects is to place dried whole leaves from the coca plant—the same one that produces the drug cocaine—into hot water and let their natural chemicals leach out into a rudimentary rough brew. I drank the tasteless tea and then ate a dinner of potatoes and some dark mystery meat from some undisclosed hoofed mammal.

Later, when I started throwing up my dinner, I assumed that the food had made me sick. It's probably just as well that I didn't

know about cerebral edema (brain swelling), which first mani-
fests itself as nausea, headaches, and vomiting. But when the nausea and vomiting didn't cease for two days, I realized that I had
altitude sickness, not food poisoning. In the mountains, the only
cure for altitude sickness is removing a climber to a lower altitude.
In La Paz, that is very difficult to do, because there is nowhere to
go. The city is surrounded by higher mountains and plains to the
west and by impenetrable forests to the east. I had to keep swill-
ing coca tea and hope for the best.

While the curator gathered permits, I stayed in the hotel room
and read the only novel I had brought along, *Elmer Gantry* by
Sinclair Lewis. By the third day, I started to feel human again and
was able to walk around the downtown a bit. I saw many beggars,
open markets where you could purchase almost anything imaginable, and a massive cathedral, San Francisco, which was built
during the colonial period, only a generation or two after Spanish
conquest of the New World. As with other parts of the old Spanish empire, La Paz struck me as desperately overpopulated and
riddled with poverty. In general, the society seemed chaotic and
on the verge of collapse at any second—kind of like New York or
any crowded capital city at rush hour, but with a heavy overlay of
hopelessness.

A day or two later, I joined the curator at an official meeting at
GEOBOL, the branch of government that oversaw exploration in
Bolivia's remote regions. There I met some government scientists,
including one who was going to join our expedition to gather
geographic data. Another addition to the trip was a Canadian
scientist, who was intent on locating a remote meteorite-impact
crater suggested by a faint outline on a satellite photo. The outline
may have been nothing more than a photographic aberration, but

it had to be confirmed or denied by sampling the ground near the supposed point of impact.

Everyone at the meeting agreed that the trip we were taking was ambitious. We were to explore a remote tributary to a major river in northern Bolivia, the Madre de Dios. This river flows northeastward into Brazil and eventually joins the Amazon, but we were going to trace the tributary upstream deep into the remote jungles of Bolivia. The people who live in remote settlements along the Madre de Dios consider themselves more akin to Brazilians than Bolivians, because their way of life as rubber tappers, subsistence horticulturalists, and gatherers is far more similar to the typical twentieth-century Amazon Basin lifestyle of Brazil than it is to the Bolivian mode of urban living near the mountainous capital.

Just as in Joseph Conrad's *Heart of Darkness* or the movie *Apocalypse Now,* our expedition was going upriver far beyond any known settlement. I knew that the plans were sketchy when they pulled out a map of the region marked with nothing but dashed lines in the vicinity of our planned trek. In general, maps with dashed lines indicate a "best guess" for locations. If the course of a river, for instance, is only dashed instead of a bold blue line, it means the mapmaker didn't want to commit to the precise location of the water channel. In the case of our map, the locations were based on very blurry Landsat images from the first generation of satellite photographs taken by the U.S. Geological Survey. There were no roads or trails in the entire northern region of Bolivia in 1986, and no mapmaker had ever even visited the region. In fact, no white men had ever ventured up this tributary as far as we were planning to go, which was one of the major reasons that the government was taking so long in granting us access.[8]

From earlier reports of missionaries, there were only small villages along the Madre de Dios, generally populated by one or a few families at most. These *barraccas*, as they were called, might be separated by many hours of river travel. A native group of hunter-gatherers also were thought to live in the region, but no official reports existed of their range. So it was possible that they might use the same tributaries as we would.

I quickly got the sense that the Bolivian government didn't have much of a grip on its northern territory. For that matter, the government didn't have much of a grip on its own capital. Bolivia in the 1960s, 1970s, and 1980s had a tradition of extremist government takeovers that pretty much fit the stereotype of a banana republic. Numerous coups preceded our expedition. My job was to collect global biodiversity data of international significance, but the expedition as a whole was approved by the Bolivian government in order to gather some very fundamental information about Bolivia. The government scientist who joined us was in charge of bringing basic geographic data back to the capital.

It took the curator about a week to get the necessary permissions from the government, but finally we were on our way to the airport to board a flight bound for the Amazon Basin. My heart raced with excitement as I thought about our upcoming wilderness experience. I was anxious to observe the intricate dependence of other species on one another and their environments. I was also excited to be in the company of more experienced scientists. I was eager to learn from them during all of the casual conversations we were going to have around the campfires under the stars of the Southern Hemisphere. This trip had everything an ambitious young naturalist could want—adventure, danger, uncertainty, and the promise of new discovery. It had the makings of a classic naturalist expedition.

Perhaps the greatest jolt of excitement I've ever had from an excursion came on that airplane flight into the Amazon Basin. There was a daily jet service from La Paz to a small but growing city far below the mountains, called Trinidad. There we would board a "puddle jumper" propeller plane to fly to Riberalta, the river town where our expedition team was gathering. Getting to Trinidad was only a half-hour flight on the jet, but it proved to be one of the most thrilling half hours of jet airplane travel imaginable. Jets leaving La Paz require runways that are much longer than normal, because the air is so thin that planes have much less lift on takeoff. About the best they can muster is a stubborn, low-trajectory release from the ground. The plane just barely seems to get off the ground before the landing gears are retracted. And then the plane does the opposite of what it is supposed to do. Instead of a rapid ascent to cruising altitude, it begins to descend. To drop from the highest capital city in the world into the basin of the Amazon, the plane cruises just slightly above the down-sloping landscape.

Within minutes of takeoff, my eyes witnessed one of the great natural wonders on the planet: the emerald blanket of endless forests coming off of the eastern slope of the Andes. From my window seat near the back of the plane, I could see monstrous trees clinging to cliffs and promontories in the clouds, high up in the rain forests. Some regions were barren, dominated by light gray patches of crystalline andesite. Soon the landscape became more uniform as we descended deeper into the basin and away from the mountains. The last ten minutes before landing brought a sight I will never forget. As far as the eye could see was nothing but a monotonously flat landscape of massive tree-crowns and impenetrable green. No people, no dwellings, no roads, no signs of civilization at all. The only things that broke up the monotony of the forest were rivers, but many of them were obscured by billowing expanses of growing vegetation. I remember feeling so insignificant in size that the forest might just eat me up. The fear of being lost to the outside world, however, only added to my sense of excitement. Somehow I was going to have to find a way to live out there where the rules of nature dominate and the cares of society have no place.

Riberalta, on the banks of the Madre de Dios, was a remote outpost in 1987. Although it boasted a population of fourteen thousand people, mostly horticulturalists, ranchers, and forest merchants, it had no roads connecting it to the rest of Bolivia. One new "superhighway"—in actuality, a dirt road—led to cities in Brazil, downstream, where the markets were larger. The locals maintained an openly jovial attitude toward flight schedules. They claimed, "Riberalta is a place you get to when you can, and you leave IF you can." They were right: our flight connection from Trinidad to Riberalta arrived thirty-six hours late!

My first day in Riberalta was filled with anticipation. There we met another scientist, a botanist from the Missouri Botanical Garden, who was going to accompany us to collect plant specimens. So the expedition consisted of me, the youngest and least experienced at tropical exploration; my boss the curator; Jim the botanist; the Canadian meteorite specialist; the Bolivian government geologist; and a local guy named Cenone, the *motorista* in charge of navigating our boat through unknown waters. Over dinner, at the Club Social, the best restaurant in a town without restaurants, we sat around the table discussing hopes and plans for the expedition. The curator said very little. The rest of us couldn't wait to load up the boat and hit the river. As I finished my meal of *pescado frita y papas*, the waiter ordered five moped taxis to take us all back to our motel. In the morning, we were to meet on the banks and head down the docks to see the newly refurbished boat that would be our home for the next six weeks.

If Mark Twain's *Life on the Mississippi* had a modern-day counterpart, it could have been written about Riberalta and its docks in 1987. There were huge, hundred-foot-long, double-decker merchant ships tied up next to tiny dugout canoes. Some boats were nearly sinking from their full cargo of burlap bags overfilled with *castañas* (the local word for Brazil nuts) on their way to markets in Brazil. Others were draped with hammocks two or three layers deep for weary mine workers on their way to or from industrial operations carved out of the jungles far from home. Canoes manned by native "Indians" of the rain forest constantly brought pelts from trapped mammals, fish for local markets, or balls of rubber the size of pumpkins, which would be bartered for sustenance.

Amid the scores of boats docking at the lively riverbank was a freshly painted, thirty-foot-long, flat-bottomed wooden vessel named *El Tigre de Los Angeles*. This was our research boat, complete with a nicely painted logo of the L.A. County Museum: a saber-toothed cat. To the locals, we must have looked ridiculous. They surely had never seen a saber-toothed cat before, and to them the boat's name translated as "the tiger of the angels." In any case, we waited another day for the supplies to be loaded, the fuel barrels to arrive, and the equipment to be brought out of storage from a local merchant's warehouses. By the next morning, it was time to depart.

The first week of travel proved to be ridiculously monotonous. The boat would hum along at a constant speed—around eight knots, more or less—from the time we broke camp around nine A.M. until sundown at six P.M. River navigation in the Amazon Basin is rarely done at night, especially by small vessels. Storms can bring down the largest rain forest trees on the banks of the rivers. Some of these trees are two hundred feet tall and ten feet in diameter at their base. As soon as they plunge into the water, they get swept toward the middle of the channel and gradually become waterlogged. Eventually, they absorb so much water that they can barely float. Most of the tree is thus invisible to watercraft, and many a boat has been sunk in Amazonian waters due to these treacherous obstacles.

Because it was the dry season, the water level was relatively low, which meant that banks of red clay and mud rose twenty to thirty feet above us. Finding camping sites in such conditions proved challenging. Without stairs or ramps, the banks of vertical mud cliffs could not be scaled, and there was no way to enter the forest from the level of the river. Sometimes we camped in the middle of the channel, where the low level of the water exposed some sandbars. These were dramatic sites but totally inert biologically, since species could not become established on a sandbar exposed for just a few months of the year. They did make for spectacular views of sunsets. But with hundreds of yards of water on either side of us, we were essentially stranded for the evening on an island far removed from the biological wonder of the tropical forest.

My favorite campsites were the *barraccas*, family settlements carved out of the forested banks high above. Usually the residents had semipermanent ramps down to the channel that ended with makeshift docks and one or two tie-offs that would keep our boat from floating away downstream. These were the only settlements we ever saw on our entire journey. Close to Riberalta, within the first two days of travel, we saw maybe fifteen to twenty such dwelling sites. By the end of the first week, we might go an entire day and only see a single *barracca*.

*Barracca*s were populated mostly by Brazilian and Bolivian families who lived on a subsistence diet. With no electricity and no running water, they usually built thatch huts and often had small gardens and maybe a banana grove. They were scantily clad in seemingly out-of-place T-shirts and shorts, obviously obtained from markets far downstream. They subsisted by collecting nuts and tapping rubber trees. Once or twice a year, they might head downstream in their dugout canoes to barter their goods for necessary items. It seemed like an extremely difficult way of life. None of them looked particularly healthy, and I believe that they represented a form of poverty that probably went unreported by the Bolivian government. We were, however, extremely grateful to the families for letting us camp at these places. On several occasions, they had unused shelters or thatch pavilions under which we were welcome to pitch our tents or throw our sleeping bags. In exchange for a good night's sleep, we left them with a parting present: canned meat from Riberalta. The curator always brought along a huge supply of canned meat, usually little hot dogs, for trading purposes. We ate lots of beef jerky, Brazil nuts, oranges, bananas, and canned beans—typical camping food.

Throughout the entire week of nonstop travel, the curator barely spoke. He obsessed over the satellite imagery, ticking off marks for every meander we traversed. The Canadian scientist, a shy man in his forties, had a disposition best described as distant.

He was as silent as the curator, merely observing the passing forest. About five days into the trip, the Canadian scientist finally voiced some of his thoughts. He said, "Only a few times in my career have I gotten myself into a situation where I looked around me and said, 'What the fuck am I doing here?' This is one of those times." The government scientist and the *motorista* spoke Spanish between themselves and otherwise were as silent as everyone else on board.

I began to feel terribly lonely. My expectation of comradery, teamwork, and discourse with fellow scientists was turning out to be a huge letdown. If this expedition was any indication of how scientists behaved in the professional ranks, it was sadly disappointing. I missed the after-dinner mirth-making with my undergraduate peers on our camping field expeditions. Science can be intensely social. The stereotype of lonely scientists working by themselves in the lab is largely wrong. But without the weekly seminars, office hours, and Friday happy hours with students and faculty, a life in science can be downright dismal. What good is knowledge if there is no one to share it with?

The hours of silent pondering I spent watching the riverbank go by also caused me to long for music. When I left, the band was pretty much dormant. I thought that devoting my efforts to a scientific expedition would be a suitable replacement. But the icy silence emanating from the scientists around me led me to reassess the emotional importance of music in my life. I had brought a Sony Walkman tape player and two mix tapes along on the trip, four hours of music in total. Those tapes only had three or four albums recorded on them, but I never got tired of listening to them. In the absence of much human companionship, those tapes and Elmer Gantry were my best friends.

As the *barraccas* became more and more infrequent sights, we turned up an extremely narrow tributary channel of a small river system called the Manuripi. If a search party were to come looking for us by flying over the region, they would have surely missed us, because the forest canopy overhead spanned the entire width of the stream. The overhanging vegetation allowed hardly any sun to fall on the shallow stream, and the banks were no higher than a meter. Soon the only way to make it farther upstream was by canoe, and our expedition hadn't brought one. We had traveled for eight straight days, and I had done hardly any collecting at all.

We tied *El Tigre de Los Angeles* to a tree near a rudimentary clearing. Humans must have visited this area in recent years, but the clearing was so overgrown with new shrubs and vines that we decided it had been abandoned. The rain forest was incredibly dense. It harbored an immense abundance of shrubs and thickets below a mature canopy of much taller trees. It's the most difficult type of jungle to navigate. It can be so dense, monotonous, and still that one feels trapped in an unwelcoming verdant maze.

I spent the first day setting up my workstation. It consisted of a simple foldout desk and folding chair surrounded by supply bins. Supplies included cotton for stuffing skins, needles and thread for suturing, scalpels and scissors for dissection, wire and identification tags for specimens, drafting pens, loose-leaf journal-log pages, jars of formalin (diluted formaldehyde) for pickling collected specimens I didn't have time to skin, syringes for injecting sodium pentothal into trapped animals for euthanasia, cornmeal and salt for absorbing blood residue from mammal pelts, and plenty of gauze to soak up excess body fluids.

Each morning, I headed out into the field to check traps and "mist nets" that I had set the night before. Small mammal snap-traps used in the field are no different than the ones used in attics and crawl spaces. The key to success, as in homes, is careful trap placement. I laid out about fifty traps in the forest and had a success ratio of around 10 percent each night. That meant each morning I could hope to collect around five dead rodents to bring back for skinning or pickling.

Mist nets are made of a mesh so fine that they appear invisible to unsuspecting flying animals. Hung loose and floppy between adjacent tree trunks, they ruffle in the wind like a huge cobweb. Small birds by day, and bats by night, get tangled in the net's fine mesh pockets and can't escape. Once caught, the animals have to be euthanized, usually by pressing their sternum and preventing them from breathing. Death is swift and humane.

Larger mammals and most birds had to be shot rather than trapped, so, immediately after checking my traps and nets, I spent the rest of the morning silently hunting with a gun that could discharge either a .22-gauge bullet or a spray of bird shot. For larger prey, the expedition also supplied me with a 20/20 shotgun.

One day I came to an abandoned *barracca* in a small forest clearing less than five hundred yards from the stream bank. It seemed like a normal family dwelling—two connected grass huts, a kitchen area that still contained some grains in a bin, and a sleeping room. Since it seemed to be abandoned, I took the opportunity to set up a few traps around the kitchen area to see what might be visiting during the night. On the way back to our camp, I found a grisly specimen too large to pickle and too decayed to skin. It was the head of a tapir, a large tropical relative of elephants found in South America. Who would cut off the head of a tapir? For some reason, it didn't dawn on me that here was relatively fresh evidence that other humans had been in this area not so long ago.

After four or five days of skinning and collecting, I had amassed a good collection of rare species. I didn't spend much time identifying them. That kind of taxonomic precision is usually done back in the museum, where comparative specimens might be consulted. But my journals were filling up with descriptions of interesting birds, mammals, and even a few reptiles that wandered too close to the traps.

One particular afternoon, it was raining so hard that I moved my work to the boat. The other members of the expedition team had taken off for a three-day camping trip, cutting their way through the thick floodplain forest in search of the elusive impact crater. The *motorista,* botanist Jim, and I stayed back at the camp.

As I was working under the canopy on the boat's rear deck, the most eerie sight floated around the bend in the stream. Through a rain forest fog of stormy, humid, midday darkness, two men appeared, naked except for loincloths of leather, standing erect near the bow of a massive thirty-foot-long dugout canoe. They both carried rifles that looked like Civil War–era muskets. Traveling no faster than the turbid current, they floated directly to the back deck of our boat. They boarded *El Tigre de Los Angeles* as if they didn't need permission. I waved and said, *"Hola! Me llamo Gregorio,"* to which they responded, *"Missionarios?"* I shook my head and quickly retreated to the opposite end of the boat, where botanist Jim was pressing leaves. Luckily, the *motorista* had noticed our visitors and started conversing with them in some kind of local tongue that I didn't recognize.

Later, our *motorista* told us that he could only converse with the men in a native language that was common to some of the upper reaches of the Madre de Dios drainage. He told us that our visitors had never seen white people before but had heard stories about white people who called themselves missionaries. They lived in small villages another two days' travel upstream. But it turned out that we had pitched our expedition tent right under a huge cache of dried meat that these hunters had hung in a tree just a few weeks prior. Our *motorista* explained that they had a good attitude about us, especially when they heard that we were only catching small birds and mammals. He showed them our guns and they had a laugh, knowing that we had no way to kill large game, such as tapir, which seemed to be the species they favored.

After about an hour, the natives headed back upriver, having secured a cache of dried meat in addition to six cans of mini-wieners from our larder. I breathed a sigh of relief as they disappeared. Here were some people whose only experience with white people was from tribal mythology about invasive and questionable social enterprises. With all the stories of diseases, land disputes, and exploitation brought by the white man, I couldn't have blamed them for wanting us dead. And since I was the only one who was using guns to take animals from their traditional hunting grounds, their choice for a sacrificial white person surely would have been me. Such reasoning seems silly now, but at the time it caused me an afternoon of dire stress.

I had a more rational reason to be scared on that day. We found out from the native hunters, through translation from our *motorista,* that the *barracca* nearby was only recently abandoned. Only months earlier, every member of the six-person clan had died of hemorrhagic fever, a grisly infection that causes internal bleeding of vital organs. The fever is spread by rodents—the very same ones, most likely, that I was so eager to trap and kill.

As the days and weeks went on, I became more and more disillusioned and lonely. The only part of the day I cherished was my free time before bed. Usually I turned in early to read *Elmer Gantry* and listen to music on my tape player. Escaping into my headphones, I fell asleep each night in an isolated state of musical bliss. Immersed in the creativity of nature each day, I somehow felt an enhanced affinity with the creativity of music. I had developed no friendships or bonds with any of the expedition team members, and they seemed perfectly content to keep their distance from one another and from me. My music, however, exerted a profoundly calming effect. One of the recordings on my tapes was a three-song studio project that I had recorded with Bad Religion. The songs never made it onto our album because we felt that they needed more refinement. I listened to those songs over and over again. I felt a great sense of comfort as I dreamed of ways to improve the band and my songwriting. Music got me through those dark, solitary nights in the forest. It reminded me that brighter times were waiting for me upon my return. After all, the band was still intact, in name at least. We never officially broke up, and opportunities for weekend gigs still existed at some of the divey Hollywood nightspots. Thinking about songwriting and performing helped put my current situation into a more hopeful perspective.

I was to enter graduate school in a few months, but that was so far off, all I could think about was getting through this expedition with my sanity intact. I was sure that I could study something interesting for my master's degree, and I was equally confident that the band could record a new album, better than any we had produced before. Nature study alone could never sustain me. Especially if it meant having to put up with stodgy, uncommunicative field scientists for the rest of my life. I needed that connection with emotion and people that music provides. It was at that remote redoubt on the Manuripi that I was able to revisit my values, reconnect with my essential self, and determine to redouble my efforts to keep the band going.

+

About two weeks after the visit by the native hunters, the expedition headed back downriver to restock in Riberalta. Although it took us eight days to travel upriver, it was only a three-day trek back to town. It felt great to have a good shower and a meal at the Club Social. At breakfast the next morning, the curator informed me that he was going to leave the expedition to attend some conferences in other countries for the next four weeks. The Canadian and the government scientist were also bailing out on the expedition. They had had no success finding evidence for the meteorite-impact crater, which meant the end of their investigations. The *motorista*, Jim the botanist, and I were directed to travel downriver to the *barracca* of a well-known local farmer, set up shop on his banana field, and spend the next few weeks collecting specimens.

I saw the curator's departure as an abdication of his accountability. I had very little confidence that I could bring the expedition to a successful completion. I wasn't hired with that sort of responsibility in the job description. I didn't understand all the implications of what it meant to be a captain of a boat, and Jim the botanist had his own agenda for collecting specimens. When we arrived at our new destination downriver, we barely saw each other. Leaderless and aimless, we sullenly went through the motions of scientific research. My nerves were frazzled. I knew that this leg of the journey was nothing more than a consolation prize.

Like the elimination round in a sports tournament, the leaders went off to new challenges, while we were sent away to kill time and look busy.

One night Jim was listening to a shortwave radio broadcast in Spanish from La Paz. He was fluent in the language, so he could make out the serious tone of the reporter, despite the extreme amount of static on the radio. Apparently, while we were going about our daily routines in uncharted parts of the country, Bolivia was undergoing another coup. It appeared to be a peaceful takeover by a right-wing group staunchly opposed to outsiders who wanted to exploit the natural resources that rightly belonged to the Bolivian people. "What this means," Jim exclaimed, "is that our collecting permits, issued by the previous government, are no longer valid!"

This news was all it took for us to hightail it back to Riberalta. We hid our specimens—which essentially became illegal federal contraband overnight—in the warehouse of the local merchant we knew. Then we began looking for any possible way out of the country. Jim was an employee of a well-established institution, so he had no problem finding a way out. His employer could simply wire money to the local airline carrier, and he, along with his specimens, would be out on the next flight.

I, however, had only one contact from my museum—the curator who hired me—and he was off in some undisclosed South American city attending academic conferences. I pictured myself getting escorted by federal agents to the nearest detention center for illegally collecting natural resources while stories of my imprisonment circulated among the academics at their cocktail parties back home. Fortunately, the merchant who agreed to store our specimens also knew of a man who had a two-seat Cessna airplane. I thought perhaps he could get me to the nearest airport with commercial airlines—Trinidad, a one-hour flight from Riberalta. I shall never forget my feeling of relief when I first met the pilot. He spoke with a Midwestern American dialect, just like the accents I grew up with. He was a missionary with a plane, and his first words to me were, "I hear you wanna get over to Trini." His airplane hangar was a makeshift grass hut in the back of some landowner's field near the municipal airstrip. He made small talk about his Indiana roots and his family's missionary work among the "natives of the Madre de Dios." He had never visited the upper reaches where we had been. But I told him about some hunters up there who were expecting him to make an appearance.

On the instrument panel of his single-engine plane, which he had flown all the way from Indiana, was a sticker that read GOD IS MY CO-PILOT. Not a word was spoken about the coup d'état, mostly because I was worried that he might not take me where I wanted to go. Who wants to be associated with a fugitive? As we taxied out to the dirt runway, my heart was racing. The engine was very loud, so conversation ceased. I was left to ponder all that had happened. My expedition had been a great adventure. But, in actuality, the expedition had ended for me weeks earlier. My enthusiasm for science was based on my belief in a social environment of mutual support and enthusiasm among all scientists. When it appeared to me that every member of this expedition was simply looking out for number one, I lost all motivation. In fact, it didn't matter if there was a coup d'état or not. I was leaving the jungle because, in a total vacuum of comradery, the expedition had been passed down to me, its lowest and—at age twenty-two—most inexperienced member. And I had lost faith in it.

The pilot flew barely a thousand feet above the canopy of the forest. He left the cockpit windows open so I could shoot some photographs. The views seemed even more spectacular to me now that I had spent weeks within that green carpet of vegetation. An endless sea of foliage spread out in every direction. But, at this lower altitude, I was given a privileged view of life's most creative landscape. The number of tree species I saw was staggering, higher than anywhere else on the planet. If we were to fly over mature forests in New York or Wisconsin, I might see fifteen or twenty dominant species per hectare. But here, there were seventy to eighty species of tree within a single hectare! And each of those trees housed an entire community of animals and plants peculiar only to that species. Thousands of insects, many of them still uncataloged by science, lived on each tree. I could have spent my entire life in a single patch of this forest and never been able to adequately describe its biological diversity. The creativity and exuberance of the life I experienced in that jungle was unimaginable.

I believed in science for the same reason I believed in music. Both bring people together, and both are founded on creativity. True, writing music, practicing songs, and listening can be lonely, utilitarian pursuits at times, every bit as reclusive as studying a tome on natural history or documenting new species in a patch of jungle wilderness. And the expedition had shown me that science can be equally lonely when scientists act like the secretive members of my failed expedition. But I also knew that biological research produced knowledge that we should not keep to ourselves. It had a story to tell. It could help people make sense of their lives.

I never again saw the research specimens that I had collected and prepared for the museum. They were stashed away in a mahogany crate at some dirt-floored warehouse in a remote Amazonian town. For all I know, they are still there today. Or, more likely, they were shipped to La Paz, where some museum scientist replaced all of my specimen tags with those containing his own signature. But all my hard work in the jungle had an unexpected consequence: it strengthened my resolve to interact with other people. Whether with scientists or musicians, I was, from then on, always a collaborator, always eager to bring discoveries from the field, as well as those from the recording studio, to groups of inquisitive people, be they students or music lovers. The expedition taught me that without a sense of collaboration, all enterprises are doomed to fail.

There would be more expeditions to come, musical as well as scientific. But as I flew over the remote jungles of northern Bolivia, I vowed never again to take part in a collective enterprise unless its participants showed a mutual spirit of cooperation and a commitment to some greater social good. In science, as in religion, one can become indentured to self-appointed authoritarian leaders with ulterior motives. I knew that it didn't have to be this way. I still had faith that a life as a naturalist could be rewarding.

But as the last rain forest tree disappeared from view, I wasn't thinking about science. The sound of a harmonious chorus and distorted guitars filled my head. I was back between my headphones, imagining a new direction for Bad Religion.

CHAPTER 7
WHERE FAITH BELONGS

You've got to choose your heroes, choose 'em well. They
could be leading you straight into hell.
—Todd Rundgren[1]

Three passions, simple but overwhelmingly strong, have
governed my life: the longing for love, the search for
knowledge, and unbearable pity for the suffering of mankind.
—Bertrand Russell[2]

All songwriters strive to tap into a universal sentiment—that combination of words and notes that will make women "weep instantly when they hear it," in the words of Nigel Tufnel.[3] But to what extent do any of us have access to that universal sentiment—the thoughts and feelings of not just ourselves but also of others? This is a question for philosophers as well as songwriters. They have debated and framed the issue for many hundreds of years and have arrived at largely pessimistic conclusions.[4] We are locked within our own skulls, some claim, powerless to draw any more than weak analogies between our feelings and the feelings of other people.

My experiences in music and evolutionary biology lead me to conclude otherwise. We may not be able to think another person's thoughts or know exactly what a person is thinking, but we have the ability to feel joy, fear, surprise, and even love so deeply that the usual distance between our experiences vanishes. In the best concerts, there is no barrier between a performer and the audience. Emotions flow both ways, as if the musicians and the listeners were locked in an intense conversation. That is one reason why most musicians are so eager to perform, despite all the hassles of being on the road. Performing provides an instant response that cannot be achieved when writing a song or recording in a studio. When a singer at a punk-rock concert throws himself on top of the outstretched hands of an ecstatic audience, the act symbolizes the bond of not only trust but common feeling that has been established between them. I've rarely jumped off stage since I "retired from the pit" early in my singing career. But I distinctly remember the feeling of synergy that overcame me in my youthful performances. When I jumped, I trusted that the audience wouldn't let me fall.

The slam pit (also referred to as "the mosh pit," or simply "the pit"), which forms in front of many punk-rock stages, embodies the solidarity of fans reacting to the intense music of a concert. To people unfamiliar with punk music, the pit appears to be a scene of unbridled aggression. Kids run wildly and randomly into each other, bouncing off each other as if they were senseless automatons. Sometimes they run in gigantic circles (also referred to as "Turkish circles"), human maelstroms that from the stage can look like spinning wheels of fire. Most pits seem totally anarchic, as if only the biggest and most inebriated participants were destined to emerge unscathed. But, in fact, there are many unspoken rules in a slam pit to keep people from getting hurt.[5] The most important is that punk fans are obliged to look out for one another. If someone falls, the nearby people in the pit are required to stop slamming and help that person get up. When girls enter the pit, they are not to be groped or assaulted. Kicking and punching are not cool, though the flailing of arms and legs sometimes can resemble kicking and punching. Punk culture has always been defiant and aggressive. But it also has been cooperative and egalitarian, at its best. Sometimes the rules of the pit are violated by jerks, drunks, and assholes. But almost everyone is aware of the rules. They emerged spontaneously from the culture of punk and from the social networks functioning in front of the stage. They are meant to establish social cohesion even as they allow individuals to express themselves as independent and autonomous agents.

Evolutionary biology has taught us many of the same lessons on a much broader scale. We are evolved social organisms that have emerged from the careless void of primitive animal relations. I may not know exactly what you are thinking at any given moment, but I know that we share a biological heritage that extends billions of years into the past, and because of this, we see the world in similar ways, experience cold and hunger and pain using the same biological mechanisms, and feel emotions that our nonhuman ancestors felt, even if those emotions are refracted through our unique human consciousness. And these common experiences and feelings bind us to one another as tightly as the onlookers are bound to the performers at a good concert.

The commonality I've been describing is also known as empathy—the ability to comprehend the thoughts and feelings of another person due to common experience. I distinguish empathy from sympathy, which is a different emotion—a sense of compassion for another person's plight. And both differ from pity, which I think of as a mixture of concern and condescension. Empathy has both a cognitive component and an affective one. We can apprehend what another person is thinking, as you do when you are reading this book. We can also share in that person's feelings, whether through art, comradeship, or love.

Not everyone feels empathy to the same degree. On the one hand, some autistic people appear to be born with a neurological condition that severely limits their ability to appreciate the emotional state of other humans, despite having similar experiences. On the other hand, sociopaths either feel no empathy or have become so adept at suppressing it that they never bother to assume another's perspective. And all of us can become so tired, frustrated, angry, or bored that we ignore our empathic impulses, even when doing so makes others and ourselves miserable.

Where does empathy come from? Obviously, almost everyone is born with the potential to exercise it. Infants just a year old can express concern for the feelings or desires of others.[6] Even other animal species, such as chimpanzees, seem capable of empathy.[7] One need only watch a TV documentary showing the gentle care displayed by such "beasts" as bears or crocodiles, as they nurture their offspring, to realize that some sort of empathic drive must be behind such behavior.

The expression of empathy in humans requires that individuals have the proper experiences growing up. If children never witness adults behaving with respect toward others, they are unlikely to learn how to do so themselves. Empathy, like most human traits, arises through a combination of our biological potential and our environmental influences. For that reason, groups of people can show wide variation in their empathic style, and the expression of it within any single group can change significantly over time.

With some exceptions, Western religions have not emphasized empathy. They are prescriptive. They impose codes of behavior based on injunctions from supreme authority, not based on the give-and-take of human interactions. Western religions define proper behavior by analogizing human nature with the behavior of mythological figures who have supernatural powers. For example, people are supposed to behave like the saints or Jesus if they want to be admired members of a Christian community. Codes of conduct, therefore, emerge from the supernatural realm and are not to be questioned by mere mortals.

Science, in contrast, is based solidly on empathy. It posits a shared experience of the world, because, otherwise, how could we agree on explanations for, and verification of, natural phenomena? More deeply, it finds and celebrates in humans a capacity to learn about the world and share our experiences through reason, logic, language, music, and art.

The capacity for empathy enables us to organize our societies in beneficial ways. Because we can see at least some aspects of ourselves in one another, we can derive ways of acting that are good for us and for society as a whole. But in order for this to occur, we have to be open to accepting other people's experiences as equally valid to our own. This is simply impossible if prescriptive codes are too strictly enforced, particularly if those codes are underlain by the unverifiable "truths" of the supernatural realm. Empathy is the best basis for human ethics that we have. It provides a solid foundation for strong personal relationships and a productive society.

+

I wasn't thinking about any of this when I returned from my aborted expedition to Bolivia, even though my experiences there showed me what can happen when empathy is lacking. I just wanted to make Bad Religion the best band I could while learning how to work in science collaboratively. I began graduate school at UCLA in the fall of 1987, but I also began writing songs with renewed fervor. I am always working on ideas for songs, no matter what else is going on in my life. I'm never far from my notebook or sketchpad. Whenever something strikes me as a possible song title or concept, it gets entered on a page of its own. I then gradually fill in those pages with lyrics or just miscellaneous thoughts. By the time I returned from Bolivia, I had many pages overflowing with ideas. But getting those ideas recorded into songs on tape was going to take time and persistence. I wasn't even sure if any of my bandmates still were interested in playing music.

Greg Hetson, who had been constantly touring with his band the Circle Jerks, was my first call. I asked him if he had any desire to play Bad Religion songs. He was enthusiastic, as always, but he added, "I'm not sure that I can devote full-time to it because the Jerks have a lot of shows this fall."

My next call was to Brett. He was preoccupied with building a new studio and was working late hours as an audio engineer to attract bands to his growing music label. But, to my surprise, he, too, was thinking of making a new Bad Religion album, and he immediately invited me to see his new studio in Hollywood. As it turned out, Jay was also hanging out at the studio quite a bit, so it was no problem to get him excited about recording some new songs.

At that time, I was living near UCLA, and the trek to the studio was about five miles. I had a beat-up old 1979 Honda Civic—the classic kind that looked like circus clown cars—and my girlfriend (and future first wife) Greta got me a job as a "salad bar host" at the restaurant where she was a waitress, so I could pay to keep my car running. Greta made good money on tips. I, however, was a food-service-industry slacker. I spent more time flirting with the waitresses and talking evolution and philosophy with the bartenders than I did worrying about the presentation of the salad bar. But I did an adequate job of cleaning up and wiping down, so I didn't get fired. Still, I failed to graduate up even to busboy, which is a good indication of all the time I spent thinking about graduate school work or longing to head into Hollywood to hang out at Brett's studio.

On any given night, I might pop in to visit Brett. Often it was after a busy day of classes and studying at the university, or after getting off from work at the restaurant. No matter the hour or day of the week, there were always two constants: Brett behind the mixing console, and some dubious character in the recording booth. Brett was taking sound recording more seriously than ever before, and I was inspired by his enthusiasm to play me whatever he recorded that day. Sometimes he would play for me atrocious assaults on musical sensibility. I remember one punk singer taking several hours to perfect the monotonous refrain "I just can't hate enough" over and over again. I was worried that Brett might be losing his sense of good songs. But through it all he was thinking carefully of ways to improve the next Bad Religion record.

Brett had given his studio the appropriately nebulous name Westbeach. For some reason, recording studios always have obscure names: Rumbo Recorders, Electric Lady, Ocean Way, Sound City, NRG, Track Record, and so on. None of the names has anything to do with the geography or ownership of the facilities. I think it's an effort to assure artists that their anonymity will be preserved when they are working. And, in 1987, as we were working on new ideas for a Bad Religion album, we had all the obscurity any artist could desire. Our L.A. scene was effectively dead. I didn't even know if we had any fans left. Still, Brett spent most of his time recording young bands at Westbeach, and Greg Hetson was touring the country visiting small and large cities with growing punk scenes. I was confident that people were eager for a reinvigorated Bad Religion.

Westbeach wasn't much of a commercial facility. It had been constructed in a tiny five-room bungalow in one of the seedier parts of Hollywood. An industrial operation sat on one side of the studio; a parking lot leading to Hollywood Boulevard lay on the other. Often a recording session would have to be interrupted because noise from heavy machinery at the next-door neighbor's business was bleeding into the sensitive microphones. In general, however, the little house was well suited for recording. Each room was soundproofed to some degree. The living room was the drum room. The front entry hall was the vocal chamber. The kitchen was a lobby of sorts. The amplifiers could scream as loud as necessary in the single bedroom. And the control room was a hall connecting the kitchen with the living room, just long enough to fit the twenty-four-channel mixing console. One of the quaint drawbacks of Westbeach was the complete lack of communication possible between the different musicians. The drummer had no way of seeing the bass player who couldn't see any guitarists or the singer, who was isolated in the front hall. Since Westbeach was a rented house, Brett and his partner, Donnell, never had the authority to smash down any of the walls and install the glass studio panes that typify more upscale facilities. Communication was audio only. All the musicians had to wear their headphones in order to hear directions from the control room. Hand signals were useless. If someone wasn't ready when the "stand by" command came from the engineer, he either had to use some sort of a talkback microphone or, more often, barge into the next room and wave his hands while jumping up and down to make visual contact with the drummer. When the drummer stopped, all the other musicians knew to stop playing. Nonetheless, Westbeach had excellent ambience and electronics, and since it had Brett's personal touches all around, it felt like a creative home to me and Bad Religion.

Since Brett was so focused on recording, playing live hadn't crossed his mind much. But in September 1987, an offer came to play at a budding scenester hangout in Berkeley called Gilman Street. Greg Hetson had other commitments with the Circle Jerks, and since we had no other guitarist, Brett agreed to play this one show. (That was the trip where we figured out how long it would take to count to a million.) It marked the first time in more than four years that the original lineup took the stage. The concert was a huge success. Gilman Street was sold out, and we couldn't believe what a crazed reception the San Francisco Bay Area punkers gave us. On returning to L.A., we all decided that both Greg Hetson and Brett should be in the band full-time and that, henceforth, Bad Religion would be a five-piece.

Brett and I immediately set out to write a collection of songs. Sometimes he came over to my West Side apartment. Whenever we had free time, we wrote songs independently. When we got together, we played each other our song ideas, usually on acoustic guitars, and often we would work out harmonies. Brett and I were fans of Simon and Garfunkel and also the Everly Brothers, so we mimicked their style and crafted punk harmonies without amplified instruments.

Within two months, we had enough songs to bring the band together for evening rehearsals. We tried to practice at least three times a week when Brett and I could get away from our work. The band members were so excited by the prospect of the new material that they put their heart and soul into every rehearsal.

There may never have been a more optimistic time for us. I was on a new journey—graduate school—and I found that delving into areas of philosophical inquiry and intellectual challenge greatly enhanced the conceptual quality of my songwriting. I wanted my newly discovered concepts and words to be audible, so I took great pains to be more eloquent and articulate when I sang. I invested in a small desktop porta-studio, bought with my salad-bar-host money, and practiced my vocal skills in my burgeoning home-recording setup. This is where I truly perfected the art of harmonies and arrangements. Brett was creating sounds in the studio with far more skill than any of our previous engineers or producers. He had completed a full-blown audio-engineering degree and was serious about using that knowledge to make the band sound better than ever before in the studio. And he, too, was being challenged intellectually by the administration of his new record label and by his budding and voracious love of literature. The entire band was invigorated by a newfound intellectual activity. Our success at the Gilman Street concert made us aware of an audience "out there" waiting to be woken up. We set a recording date, and, as the date got closer, our rehearsals became more frequent and longer.

Even though the rain forest expedition was behind me, I never let go of the deep emotions that I experienced in Bolivia. I remembered the long days and nights filled with loneliness. The miserable stillness of tropical heat is nearly undescribable to those who are only accustomed to the dry winds of Southern California. In the jungle, a constant blanket of dampness enshrouds you like a down comforter. It only gets worse as the layer of wetness pools on your skin. Yet I also felt privileged to have worked in one of the last true terrestrial wilderness areas left on the planet. I remembered thinking that, despite the misery I was experiencing, life's secrets were more likely to be revealed in the forest than in the concrete jungle of some urban metropolis.

In one of my field notebooks, I wrote, "The businessman, whose master plan controls the world each day, is blind to indications of his species' slow decay." I proposed this as the lyrical underpinning of a song that became the title track of our recording project. The tone of the other lyrics revealed our simultaneous disappointments with the way the punk scene had died out years before and our hope for a more enlightened outlook that we felt the genre was ready to embrace. Brett and I decided that we would call the album *Suffer*.

At our rehearsals, the band played with such cohesion and ferocity that we moved the recording date forward by a few days to capitalize on our enthusiasm. Sometimes too much rehearsal can take the life out of a performance. We wanted to get the basic tracks recorded while they were fresh and raw. I'll never forget the excitement I felt on the first day of the *Suffer* recording sessions. Brett was more careful than ever in setting up the microphones. His trusty studio partner, Donnell, spent hours moving microphones very slightly in front of a drum head or speaker cabinet while Brett listened to barely audible variations in timbre or tone, until they found the perfect location. This was all new to me. I learned a lot about audio engineering from watching them. At the same time, I couldn't wait to start playing the songs.

After hours of tedious individual auditioning, the band started playing together. I couldn't believe what I heard coming through my headphones. The clarity of the instruments, the separation and space created by the stereo imaging, and finally the crispness of my own voice were unlike anything I had ever heard before. The first song we recorded was "Land of Competition," my homage to returning to Los Angeles. At first, I couldn't sing, because I was too busy smiling and even laughing out loud from pure auditory joy. Everyone was isolated while we were laying down the basic tracks, but when we went into the control room to hear the playback, Brett and Donnell were grinning like madmen. Nothing can compare to the overbearing sense of awesomeness when a superb piece of audio is created in the studio. It's a joy like no other. The recording process is part science, part performance, and part sheer luck. We still joke about it every time we enter the studio: it's "planned spontaneity."

The *Suffer* sessions were absolute exhilaration. No drugs were necessary to keep us going—we were high all the time on pure excitement. Brett was producing sounds that none of us had heard ever before on punk records, and I was producing harmonies that seemed to be made up on the spot. Each band member played as if he were discovering something new about himself. It was as if we were releasing all the pent-up creativity sequestered by years of inactivity. Our sessions went all night. We slept during the day and reconvened around noon for another round of recording. It was a nonstop creative outpouring that took all of our collective intellect and ambition. In seven days, the entire album was recorded and mixed, a remarkable achievement.

That album became known around the world and spawned a string of future albums and tours. *Suffer* is often credited with reviving the dormant punk scene in Southern California. Many punk musicians who later achieved worldwide acclaim, including many of the leaders of the "grunge" movement in the early 1990s, have said that they were greatly influenced by that album and the ones that followed. *Suffer* was chosen as the album of the year in 1988 by two of the most influential punk magazines of the time, *Flipside* and *Maximum Rock'N'Roll,* both of which printed Jerry Mahoney's painting of a punk teenager encircled by flames in his suburban wasteland. It was a tremendous creative and commercial success. But, as I've learned from my study of evolution, success can bring unexpected pitfalls.

+

As I raced back and forth along Sunset Boulevard from UCLA to Hollywood in those days, my mind was constantly drifting. I had become the head teaching assistant for a traditional comparative anatomy class, which meant I was in charge of teaching a lot of premed students that understanding evolution is important even for doctors. I enjoyed the interactions with my premeds, because they were generally eager to see how tissues and organs of "lower" animals could be relevant to their favorite animal, human beings. Their inquisitiveness forced me to bone up on classical embryology and study all sorts of historical evolutionary explanations. I was learning an incredible amount, because as my adviser told me, "You never really learn a subject until you have to teach it."

Invariably, I used music as an escape from the tedium of academics. Whenever I got too restless in the rare books room at the biomed library late in the afternoons, I would scuttle through the turns and sudden stoplights of Sunset Boulevard to visit Brett at the studio. After the release of *Suffer,* we wasted no time in writing and recording another album, *No Control,* and a third after that, called *Against the Grain.* We released those three albums in two and a half years, and each was more refined than the one before. On almost any day I could escape from school and get my fill of studio activity. Sometimes other bands would be recording. Other times, Brett would play me new ideas for songs. Often, I had tapes to play for him from my home porta-studio. We also had a rehearsal space for the full band, which led to hours of evening pleasure while we were working on new songs. Every morning, however, I was back on campus taking my classes, tending to my teaching duties, and researching topics for my master's degree.

A lot of people I have known spin themselves in circles by vacillating between various projects without ever moving forward (my friend Ron calls it "proj-ing," always drafting plans but never carrying them out). But the back-and-forth between music and academics suited me well, and it created a lifelong routine that I still practice today. Whenever one field gets too tedious, I jump headlong into the other. If I hadn't made progress in both areas, I could easily have been accused of being an irresponsible dilettante. But I continued to up the ante in my studies as well as in music, though I probably seemed scattered and unable to sit still. During the three years that I was earning a master's degree at UCLA, I completed three years of summer fieldwork in the mountains of Colorado; wrote a thesis on the earliest vertebrate environment; taught nine classes on comparative anatomy, evolution, and paleontology; recorded three albums; and toured with the band throughout the United States and in Europe.

This entailed a lot of schedule-juggling. UCLA is on the quarter system, which means the summers drag on and school doesn't start again until October. I was able to do fieldwork for a month from late May to late June and still have time to tour with the band during the summers before school resumed. The year 1988 was particularly hectic. I remember my future mother-in-law saying to me during a school break, "Can't you just sit still and enjoy this lovely afternoon?" I think she was trying to suggest that I devote a little more time and attention to her daughter, but I was too unaware and insensitive to understand it at the time. All I focused on was getting outdoors to go hike the awesome cliffs nearby. My future mother-in-law lived in "north county," San Diego, only a mile from the fossiliferous Del Mar Mudstone and one of the most spectacular exposures of the Torrey Sandstone, right along the beach. It's true that I was far more interested in fossils than I was in spending a quiet afternoon sunbathing or sitting around making small talk. Nonetheless, I got married to Greta in the summer of 1988. And, within a month, Bad Religion set off on our first full-fledged U.S. tour.

People who marry touring musicians have to be strong, faithful, and self-confident. Ironically, those exact qualities are often in short supply among those associated with the music industry. At the time, as a twenty-one-year-old student, I was no exception. I could tell that my role as a singer/songwriter was beginning to clash with my academic ambitions and domestic life as our albums were more widely distributed and the band got more and more offers to play concerts in far-off places. But I thought I could adapt. Upon returning from our third European tour, I moved with Greta to the quiet city of Ithaca, New York, to begin a PhD program at Cornell University. No one in the band was upset that I was leaving L.A. I assured them that I would be constantly writing songs and would return whenever it was time to rehearse or record. Around that same time, Jay was planning a move to Vancouver. All of us were confident that we could pursue other interests and still be committed to Bad Religion. After all, most bands have "off seasons," like sports teams, and most bands that tour internationally have members who live in different places. But the combination of PhD work, teaching, writing music, traveling to L.A. for recording, and going on tour would take its toll.

+

The details surrounding the infamous punk riot at the El Portal Theater on December 29, 1990, are fuzzy to me. I was, after all, a bona fide L.A. expat by that time, and I was struggling hard to find my identity as a PhD student at Cornell. So I was forlorn when I got a call from my bandmates shortly before Christmas telling me that I had to return to L.A. for a show at a theater in North Hollywood. Based on the success of our European and U.S. tours over the previous two years, I was staunch in my belief that Bad Religion deserved to play in the city's premier rock venues, not just in a rented-out hall for a weekend of chaos. Nonetheless, I cut my Christmas break short and flew out to play the concert, determined to fly back to the East Coast the morning after the show to prepare for another semester of laboratory teaching and graduate study.

As was my custom, I showed up at the venue after the opening acts, Pennywise and NOFX, had taken the stage. When I arrived "backstage" (it was just a movie theater that had been rented out for one night to host the Bad Religion concert, so there wasn't a

dressing room or production area), I heard one of the members
of Pennywise yelling into the microphone. He was saying something about the cops and about how it was unfair that they were
shutting down the concert. "Shutting down the concert?" I said
to myself. "That means I don't have to be here!" I got back into
my rental car. As I rounded the corner and drove past the front
of the theater, I remember seeing some angry punks smashing
the box office windows. "How unoriginal," I thought. Because
of the punk riots I'd been involved with almost ten years before,
the sight of these punks acting out didn't faze me that December 1990 night.

Back at my hotel, I was watching the late local news before
bed when the newscaster said, "When we come back from commercial break, heavy metal band Bad Religion is involved in a
riot in North Hollywood. . . ." My astonishment at the mention
of my band on the television news was quickly overshadowed by
annoyance. "Heavy metal band!?" I thought to myself. "We've
never played heavy metal." But then I watched the news clip and
was shocked to see that what had started as a smashed window in
the ticket-office booth had become a full-blown chaotic riot with
police cars and arrests and bloody violence. Still, I went to sleep
that night without a single phone call from friends or family.

As soon as I got off the plane in Ithaca the next day, I went
to the television to see if there were any updates on CNN. To
my amazement, the Bad Religion riot had become national news.
It was a recurring story every half hour for that entire evening.
"Heavy metal band Bad Religion sold too many tickets, fire mar-
shal cancels performance, police called in, riot ensues." We had
no manager at the time, nor a press agent to handle the media attention. The story simply took on a life of its own. The promoter

who oversold the venue wasn't talking, so the news people just made up the story as they saw fit. We, of course, had nothing to do with the ticket sales, but our name was bigger than that of the promoter or the venue, so it became Bad Religion's riot.

In those days, I rarely paid attention to business affairs. I was content to focus on my academic work and knew very little about commerce or administrative affairs. I was prepared, therefore, to just let the riot story run its course and move on with my PhD. What I didn't realize was that our fans were extremely disgruntled. They believed that we had ripped them off. Since we hadn't formally commented on the riot with a press statement, they thought we had taken money from the ticket sales and run with it. It could have been damaging to our reputation and our career if we had let the story escalate.

But Brett was very savvy. He was insistent that the promoter maintain his obligation to us and our fans. We owed our fans a show, and they were holding tickets that guaranteed them a performance from us. Brett was able to see the chaotic upheaval of the riot as an opportunity to do something good not only for our fans but for the band.

Brett insisted that the promoter honor the tickets and that the show be moved to an even bigger established theater, three times the size of the makeshift El Portal. The promoter agreed, and the next day announcements were made on local radio stations that in two weeks Bad Religion would appear at the Hollywood Palladium, the mecca of venues for internationally famous touring bands.[8] It was announced that all tickets from the El Portal would be honored and that additional tickets were available.

The radio promotion had a profound effect. The 3,500-ticket capacity of the Palladium was sold out within the week. Those who held the original 1,000 tickets for the El Portal Theater felt privileged instead of ripped off, while the other 2,500 people looked forward to seeing what all the fuss was about. The sold-out concert at the Palladium established a public perception of Bad Religion that never vanished. We were seen as a legitimate headlining act that had commercial potential and could claim an important role in a growing music genre. Before the riot, we had never played on the big stage of legitimate venues except as a "go-to" opening act for more famous national bands that were on tour and passing through L.A. We had repeatedly insisted that we could sell out large venues, but no promoters were willing to take a risk on us. They saw us as overconfident kids who didn't really understand the music business. We had to be content to be the undercard while more established acts used us to open their shows and attract our large fan base.

In January 1991, that perception finally changed and has not changed since. The riot produced a short burst of unexpected calamity, and Brett capitalized on the chaos to great effect, creating a positive fallout for everyone—the band, the fans, and the promoter. Life is full of sudden, unpredictable, drastic changes separated by long periods of predictable, comfortable experience.

✦

Yet even in our most successful moments, tragedy lurks in the background, waiting to remind us of its presence. In 1994, after we'd established ourselves as the premier punk band from Los Angeles, I learned that Brett was quitting the band after fifteen years of partnership. It was a surreal moment. At the time, I was at Cornell, involved in graduate research on the evolution of bone. At the same time, I was meeting with big-name producers in New York and negotiating with major record labels. The band had just completed our major-label debut album after signing a contract with Atlantic Records and SONY International. We had offers to headline some of the biggest festivals in Europe later that summer, when our new album was released. I had recently purchased a new house outside Ithaca, and I was busy making it into a factory of productivity—complete with a home recording studio—as well as a fun and functional place for my family. I pictured it as a musical family wonderland, like a punk version of the Jackson estate, or perhaps closer to Frank Zappa's "Utility Muffin Research Kitchen." Everything pointed toward continued prosperity and success.

Yet, looking back on this period, I can see its darker side. Almost imperceptibly, my life had morphed into an incredibly complicated labyrinth of deadlines, conference calls, and commitments. The project I had begun in graduate school demanded hundreds and hundreds of hours of research with both light and electron microscopes, and I had spent two years learning to use both. A lot more was expected of me for a PhD than for a master's degree. And then, sometime during my third year of PhD work, the major-label offers came. Being affiliated with a major label meant daily phone calls with our recently hired manager, constant calls from the "creative" departments at the labels, and many other new obligations. Both SONY and Atlantic had devoted teams of employees to further the cause of Bad Religion, all of whom needed direction from the "creative element," which was essentially Brett and me.

I thought I could handle everything. My advisory committee met with me every few months, and my microscopy was done on my own time, between conference calls and trips to New York or California. Tours were kept short, except in the summertime, when school was out. Even during the summer, however, our tours might last six weeks but then we would take three weeks off for recuperation, during which time I would head back to the microscope lab to try to catch up through long hours of work.

Cracks were developing in almost every enterprise that depended on me. The band's manager was entertaining new job offers because of the perception that he had taken Bad Religion to a new level. Meanwhile, Brett and I had pretty much stopped communicating. He was very disgruntled with the way the band was evolving. Also, the label that we started together in name (Epitaph Records), which he worked so tirelessly to build into a legitimate company, finally had become hugely successful. Brett's prowess in the studio led him to sign bands such as the Offspring, NOFX, Pennywise, and Rancid, all of which had hit records, and Bad Religion records were selling like hot cakes, too. All of the business of selling and making records required a full-time commitment from Brett in his role as a label chief. Brett's life, like mine, had become immensely complicated. One night he called me and said, "Greg, I got in a fight with the other members and I'm quitting the band." He used the excuse of the fight so that he didn't have to include me in his list of reasons for quitting, but I knew that our lack of communication was a factor as well.

I thought that my world was going to end when I got that phone call from Brett. A time of great success instantly became a time of great worry. I was fantastically busy and dealing as responsibly as I could with a strained marriage and recent fatherhood. Now I had a major crisis in my creative life as well: how could Bad Religion carry on without Brett?

I never once considered quitting the band. I simply decided that I had to stick with what we had begun together. I decided to continue to live my life and put my faith in other things. I took the opportunities that presented themselves at that moment, without worrying about any larger implications. After all, our contract called for us to complete three more records. I had a strong feeling that writing and recording music were in my blood. I couldn't let Brett's departure dissuade me from doing something I loved. And I was still in the privileged position where people wanted to hear more of my music. Besides, I had a new mortgage, graduate school to attend, and a family to feed. I believed that a music career could offer me the things I needed to remain happy and be responsible at the same time.

The cracks, however, began to widen. Because touring, promotion, and songwriting were taking so much of my time, I had to put my research on hold. My advisory committee was understanding and allowed me to take a leave of absence of unspecified duration. (It was six years before I would return to working on a dissertation.) Then, Bad Religion's manager finally found his dream job and decided to quit.

Most catastrophically, not long after our second child was born in New York, Greta and I got divorced. This was a great tragedy in my life. I never regretted the energy and determination I put into my music and academics, but there were many other things I regret about those days. At the time, I didn't have the maturity to recognize the larger picture of what my absence meant or what impact my passions had on my wife. Life, in the form of opportunities and commitments, was hitting me at an alarming pace, and I believed I was doing a good job of keeping up with it. But what I believed was adaptation was actually an illusion of order. Many things were going wrong with my marriage, even though I felt that I was successfully adapting to the inevitable strains. Ever since my divorce, I have been biased against adaptationist thinking. I have come to recognize that the only way to get through life is to respect the needs of others and accept a lot of failure and imperfection along the way.

Over a very brief time—less than two years—I experienced an upheaval that changed the complexion and the direction of my life's course forever. I didn't quit being a father, of course. I had the utmost confidence that I was a good one, and I love being one, so there was no reason to stop. In fact, I think that I loved my young children so much that I was motivated to try to do even more as a father, despite the divorce and the hectic schedule. I also believed that I could do even more as a songwriter, despite Brett's absence. Whether I achieved my goals or not is somewhat immaterial. The point is, I continued to do what I loved. The kids came first, the band came second, and my education came third. By establishing and sticking to those priorities, I opened the door to the possibility that something good could come out of tragedy.

As with the previous turmoil surrounding the riot the tragic upheaval at this point of my life ended up leading to some good things professionally (and many good things in my family life as well). As the sole songwriter for the band, I knew that I would need another creative voice in the studio. The labels were very supportive and helped the band hire famous producers—Ric Ocasek and Todd Rundgren—who taught me a great deal. Furthermore, our superb guitarist Brian Baker joined the band during this tumultuous period. And it was during this period that the band experienced its greatest worldwide growth and most successful tours.

Might we have achieved this fate anyway, without all the chaotic hardships going on with band members, managers, and families? Possibly, but that seems like a moot question to me. As in evolution, "what might have been" is not relevant. "What exists now" is what we're left to hold, ponder, and relish. As so many of my friends like to say, "It is what it is," which I take to mean, "We cannot go back and change the course of history, so we had better learn to accept what we have."

✦

Many religious believers mischaracterize naturalists as people without faith, but that is absurd. Everyone must believe in something—it's part of human nature. I have no problem acknowledging that I have beliefs, though they differ from more traditional kinds of faith. Naturalists must believe, first of all, that the world is understandable and that knowledge of the world can be obtained through observation, experimentation, and verification. Most scientists don't think much about this point. They simply assume that it is true and get to work. But this assumption has relevance to people other than philosophers. When intelligent design creationists, for example, speak of replacing methodological naturalism in science classes with theistic naturalism, they are threatening to remove this assumption from the shared presuppositions of public discourse.

Most people who believe that the world is understandable and knowable also believe that evolution has taken place, with all that evolution implies—the unimaginable age of the earth, the immense amounts of suffering that have occurred to no purpose, the undirected course of evolutionary change, the evolution of humans from nonhuman ancestors. Embracing evolution is an intellectual, emotional, and moral challenge. As we've seen, evolution has an inherently anarchic element. It cannot be controlled or directed except under limited and artificial circumstances. Natural selection is a part of evolution, but it is not necessarily the central part. Good and bad luck, random changes in the environment, nonadaptive traits, the interactions of organisms with one another, and the natural creativity of matter are all prominent parts of evolution as well.

For that reason, it is wrong to dwell on fitness and adaptation as the measures of evolutionary value. If natural selection were the predominant force in life, it might be logical for someone to say, "I am more fit than you are, so only I should reproduce," or "My biological advantages mean that I am entitled to more resources than you are." An accurate view of evolution, in all its multifaceted and anarchic glory, leads to a quite different set of conclusions. We are all evolved creatures who share a common way of perceiving and responding to the world. And yet each of us is unique, the product of an irreproducible set of causal events. Given that we cannot judge people on the basis of their biology or their fitness with respect to some arbitrary criterion of optimality, we have to conclude that all human variants are equally valid. (This conclusion can be derived purely on ethical grounds as well.) None of us is advantaged because of evolution over any other, whether strong or weak, able-bodied or disabled, woman or man, black, white, or any other color. Simply by existing as part of the human species, each person automatically has an inherent worth and dignity.

One of the ways in which people try to tear down evolution is by claiming that it makes the world meaningless. They quote Steven Weinberg, who wrote in his book *Dreams of a Final Theory*, "The more we know of the cosmos, the more meaningless it appears."[9] Or they quote Richard Dawkins from *River Out of Eden: A Darwinian View of Life*, "The universe we observe has precisely the properties we should expect if there is, at bottom, no design, no purpose, and no good, nothing but blind, pitiless indifference."[10] Or they take the words of my thesis adviser, Will Provine, out of context, when he wrote, "[Evolution teaches] no ultimate purpose, no free will, no ultimate meaning." Never mind that all three of these authors have written at length elsewhere about the value inherent in life.[11] People who fear the naturalist worldview are quick to attack naturalists as uncaring reprobates.

There is a classical flaw in their reasoning. They are confusing conclusions from one domain and trying to apply them in another domain where they are not valid. I don't believe, for instance, that evolutionary biology or any scientific endeavor has much to say about the value of love. I'm sure a lot can be learned about the importance of hormones and their effects on our feelings. But do the bleak implications of evolution have any impact on the love I feel for my family? Do they make me more likely to break the law or flaunt society's expectations of me? No. It simply does not follow that human relationships are meaningless just because we live in a godless universe subject to the natural laws of biology.

Humans impart meaning and purpose to almost all aspects of life. This sense of meaning and purpose gives us a road map for how to live a good life. This guidance emerges spontaneously from the interactions of human beings living in societies and thinking together about how best to get along. It doesn't require a god or a sacred text. Furthermore, from the naturalist perspective, the knowledge that is acquired in pursuit of a good life is subject to further observation and verification. Life is a work in progress, and recognizing errors can lead to correction. We should enjoy and make the most of life, not because we are in constant fear of what might happen to us in a mythical afterlife, but because we have only one opportunity to live.

+

There is another aspect of human life in which people must have faith. All but the most unfortunate of us will be involved in loving relationships at some point in our lives. These relationships transcend our normal interactions with other people. For example, empathy is a component of love, but love is more than empathy. Love changes a person. It involves a willingness to do special things for someone, which would not be done for others. It reveals humans at their best, but I've also seen it reduce humans to their worst. It's no surprise that so many songs dwell on the exhilarations and heartbreaks of love.

Love requires faith. Evidence can be irrelevant with those we love. My feelings of love are subjective and cannot be verified by anyone else. When I met Allison, the woman I married in 2008, I immediately knew that I had an indescribably good feeling whenever I was with her. But there was no way for us to measure, experiment with, or even deeply understand our desire for each other. Love is a unique feeling and cannot be experienced except by the person who feels love. The only other person in the world capable of agreeing with me on this particular subjective experience is Allison, the one who shares the relationship with me.

Love reveals a different kind of faith than the faith in God required by most religions. It is based on a subjectivity that applies only to two people who claim to be in love. I believe that Allison loves me, even though I recognize that her love for me and my love for her are not subject to verification as empirical facts. Love is not a fact—it's an ongoing gamble. But there are ways to improve your odds, I've learned. Just as when I jumped into the pit and correctly gambled that my audience would hold me up, my faith is based partly on confidence in myself and partly on the ways in which Allison shows that our love is profoundly important to her. In that sense, I am a faithful person.

CHAPTER 8
BELIEVE WISELY

I love the diversity of the world. I feel that one species,
mankind, doesn't have the right to exterminate part of this
creation, this wonderful evolutionary development, and that
we must do our part to preserve what nature, what evolution,
has produced.

—Ernst Mayr[1]

Natural history is a very important part of my enjoyment of
the world and always has been. . . . If I see something or I
read about something that seems to be puzzling, that's the
start of a piece of science.

—John Maynard Smith[2]

There was a call from my hotel-room phone: "There's some
talent waiting for you in the lobby." "Talent" is a code word
in rock lingo. It translates to "fine-looking young ladies hope-
fully interested in sleeping with a band member or, if that fails, a
roadie." I had just returned to the hotel from our concert in Rio
de Janeiro and had arranged for a seven A.M. wake-up call. It was
nearly midnight when our tour manager alerted me about my
"guests" downstairs. I was lonesome and far from home, and Rio
women are legendary. But I had already done the groupie thing at
an earlier stage of my life. I was dreaming instead of my planned

adventure for the coming morning—seeing the last vestiges of Atlantic rain forest.

From the ages of fifteen, when Brett, Jay, and I formed Bad Religion, to nineteen, we played shows almost exclusively in only three states: California, Arizona, and Nevada. After-show sexual encounters were easy to obtain. Some were friends; many were one-night stands. Mostly it was just good, dirty fun at a time before HIV became worrisome.

"Tell them I'm not here," I said to my tour manager.

"They'll just wait all night," he responded. I knew he was right. South American fans are tenacious. They often camp out on the driveway of the hotel where we are booked and greet us as we arrive from the airport. You can pretend you have something more important to do, but they will wait you out until you take a photograph with them, sign their CDs, or, in the case of many of the women, take them to your room. How they found out which hotel we booked was a mystery to me at first. Eventually, I figured out that the hotel sales office was staffed by fans, who wasted no time in alerting their friends.

I went to bed alone that night and woke up as planned, at seven A.M., to meet a professor and his wife, both biologists from the University of Rio de Janeiro. My graduate school professor had arranged a day of discovery for me in the coastal mountains outside of the city. Colleagues of his were scientists at the university, and they enjoyed the comradery of scientific visitors. They also were eager to spend time with me, since the night before I had given them and their graduate students backstage passes to our concert. They decided to take me to their favorite field sites along the coast, and then we were going to end the day high in the coastal rain forest called Tijuca.

When I saw my guides pull up in their 4×4 Toyota, I darted for the hotel door. "Gregorio!" called a girl from behind a lobby pillar, where she had spent the night on a lobby armchair. "Why you don't come down last night?"

"I had to get sleep because today I'm going to do fieldwork in the *floresta*." Two other girls joined her and peered with confusion at the Toyota idling outside. One of the girls made me an offer, in broken English, to come with her to visit her parents' ranch in the Pantanal, a marshy region in central Brazil. For a moment, I remembered that there are more species of plants and animals in that part of Brazil than almost anywhere else on the planet. The Pantanal is difficult to reach but famous for its wildlife, and I'm sure I would have had a fun time with the girl at her remote family cabin. But I had to decline her offer. I already had an incredible day of sightseeing planned around Rio.

Even now I wonder what I missed. What kind of man in the prime of his life would turn down the advances of beautiful Brazilian women and instead head out to look at birds, trees, reptiles, and amphibians? But this particular visit was the culmination of a dream that began in high school, when I read Darwin's *The Voyage of the* Beagle. I pictured myself in the forests he described on his adventurous journey along the coast of Brazil. Today, tropical forests are disappearing fast. Any hope of recovery and conservation is going to come through dialogue between the public and people who have been privileged to visit the last vestiges of this dying wilderness. I wanted to be qualified to take part in those conversations, and that desire took precedence over my carnal urges.

+

Why should any of us care about nature? Does it really make sense? To most people, nature is a fun place to visit once in a while. Many Southern Californians think that a trip to the beach or to gaze at the redwoods is a way to commune with nature. Those in upstate New York head out to their local waterfall and consider themselves lucky to be able to witness nature in all its splendor. To them, nature is a thing, not a process or a source of creativity. They proclaim that they "care about nature," because they want it to be there when they go back.

But if nature is a thing, it can always be reconstituted after it is lost. This gives most people a false sense of reassurance that it doesn't matter if our voracious consumerism leads to the extinction of some exotic species. By considering nature as a thing or a place, the average citizen rests assured that an endangered species must have a similar sibling species living somewhere that can act as a suitable replacement in the case of extinction, or that if a habitat is destroyed, there are plenty of places just like it that aren't in danger. This sense of security is totally false. Every species is, like every individual, a totally unique outcome of unrepeatable causal events. All species and all individuals are ongoing natural processes. As with death and extinction, restarting a process, once it is eliminated, is virtually impossible.

The word "nature" doesn't really mean anything. In a manner of speaking, everything is natural. That's why, for instance, beverage companies can say that their products are made from "all-natural" ingredients even though, for sweeteners, they use manufactured corn syrup from huge industrial processing plants. What's natural about something that has to go through numerous industrial chemical processes to be used in your drink? Relax, they say, everything is natural.

I have a similar beef with the word "God." If God is every-thing and everywhere, then what purpose does the word serve? If it explains everything, it explains nothing. But if it describes something important, then it should be observable by everyone, examined, and shared with other people.

From a monist perspective, nature is something more than just a place. It is everything that can be observed, experimented with, and preserved and cared for. Some aspects of nature may still be mysterious to us, but the naturalist worldview posits that all phenomena are amenable to observation, experimentation, and eventual understanding. Natural things undergo their own development with or without us. Natural food might be culti-vated by humans, but it was made by biological processes that occur without human intervention. The plants in a forest grow in the absence or presence of human contact. Nature is a biologi-cal, geological, chemical, and physical process. We can impede it, harm it, redirect it, and contain it, but we cannot stop it, and we have less control over it than we would like to think.

I care about nature, but I care about nature in a different way than many people do. Nature to me is the source of life's unfold-ing development. This development, in turn, forms a text with a meaning that needs to be unraveled. With each visit to a natu-ral place, with each reading of a natural history account, with each study of a natural phenomenon, I find new understanding. A story becomes more intelligible, though the story has none of the elements of a traditional narrative. There is no denouement, no main character, no crisis sequence, and no ending. The story is plastic and changes direction as we learn to read more of nature's text and as more text is deciphered. Learning about nature is a lifelong quest with an almost limitless subject matter.

People are part of nature, and, of course, I love people.[3] The good ones make me feel as though we were made for one another. Even the bad ones intrigue me—how do they reconcile their actions with the web of relationships in which we're all embedded? I'm not talking here about the love I feel for my family and friends, which is so ingrained as a sense of faith that I don't question it. I'm talking about a love amenable to investigation. We are only beginning to understand the nature of human contact, good and bad, and the motivations behind human actions. My love of people helps satisfy my curiosity as a naturalist.

If there is no destiny, there is no design. There's only life and death. My goal is to learn about life by living it, not by trying to figure out a cryptic plan that the Creator had in store for me.

✛

Near my home in upstate New York, I often ride my BMW motorcycle along the narrow lanes, unpainted with lines, that wind through pastoral scenes that have changed little in the last two hundred years. Farmsteads are intermingled with abandoned fields and secondary- or tertiary-growth forests, once logged but now full of mature woods again. The highlights of my rides are the deep gorges and glens that harbor huge trees clinging to high walls of stratified shale. These canyons are too steep for harvesting timber, and their forests have remained intact for thousands of years. The distribution of trees on the slopes depends primarily on the direction from which they are hit by sunlight. The north- and east-facing slopes harbor large stands of my favorite tree, the eastern hemlock. Hemlocks are a type of pine tree. They are found in wet, cool habitats that get relatively little sunlight.

Their enormous ruddy trunks run straight for almost thirty feet before the first branches appear. Those branches bear, at their tips, millions of tiny, flat evergreen needles with gentle, rounded points. Beneath is a thick, accumulated organic carpet of old pine needles, tiny "pine" cones, and bark that renders each footstep silent as one passes.

I love these forests, as I have loved many of the forests I have been lucky enough to see during my life. I have a farm in these New York woods, where my family and I participate in the restoration ecology "movement." I placed quotation marks around that last word because it isn't a wildly popular trend in our country. Restoration ecology is a bit of an esoteric concern usually only promoted at universities, like Cornell, that have good programs in ecology.[4] The concept is simple: we've spent hundreds of years degrading the ecosystems we take for granted. Now we need to make decisions that restore those ecosystems as much as possible, with the ultimate goal of restoring the health of the biosphere. It might sound arcane or abstract. But simple decisions can add greatly to the health of the land. For instance, many of the mature trees on our farm were struggling with invasive "strangler" vines that took hold decades ago, when loggers came through and removed the old-growth trees, allowing more sunlight to reach the vines on the forest floor. When I removed the vines with industrial shears and chainsaws—an honest job of manual labor (and good exercise)—the mature trees showed dramatic improvement within a year.

Good decisions can improve even the most modest-sized plot of land. For example, most people are persuaded by commercials that their lawn needs chemicals to be healthy, so they add all sorts of nasty compounds to the soil, which destroy weeds and kill animals. Biased by a belief that clean-looking golf greens, tended by professionals, are the best possible form of lawn, these people are convinced that having just one species of grass is the optimal solution. But the more chemicals you add to your lawn, the more you destroy its ability to self-regulate pests in the soil, and the more you impair its ability to self-propagate a variety of attractive green plants. My house has a big lawn. For eighteen years, I've never added a single chemical to the soil, and I have no fewer than eight species of grass and "weeds" that look green through drought and cold times. I keep it mowed so that plants usually considered problematic (like dandelions) never are able to take over. Instead, their attractive green leaves make the lawn look richer. Likewise, without chemicals the soil is full of insects and other animals that burrow and churn the soil (a process called bioturbation), which keeps the soil well-aerated and healthy. In contrast, my neighbors' lawns often have bald patches of hard, compacted dirt during dry times and in winter.

Complex and balanced ecosystems are the rule in nature. To take another example, our bodies can be seen as elaborate ecosystems that support a plethora of interacting tissues and organs. These entities have autonomy to some degree—for instance, they can fail or hypertrophy on their own—but they also function with input from each other, as a team of sorts. Even though all of our organs and tissues have DNA with the same genetic information in their nuclei, the cells of each tissue transcribe and translate only a portion of the code. That's how dermis, for example, makes collagen, while the nasal lining makes mucus. When all our tissues and organs act together, the "self" results, an individual organism.

Other organisms live in and on each of us. Bacteria, for example, suffuse our guts, our skin, and all of the openings into and out of our bodies. They have different DNA than we do and reproduce according to their own rules. Yet they participate in making us who we are as individuals. In that respect, the self is like a forest or a lawn. It has individual components that seem to function autonomously, but each component is enmeshed in an intricate web of dependence. Some of the entities can be eliminated without totally destroying the larger web, like the removal of a kidney or an ear in a human, the elimination of elm trees from a mixed-hardwood forest, or the eradication of weeds from a lawn. However, certain elements are so vital that the larger web's existence is in jeopardy if they are removed. Think of the loss of a heart or liver in a human, the clear-cutting of all timber in a Pacific rain forest, or the compaction of soil in a newly seeded front lawn. Even when seemingly nonessential species are removed from an ecosystem from overhunting, overfishing, or "pest management," the result can be analogous to the wanton removal of an organ from our body. Sure, I can live with only one lung, but I want to do everything in my power to avoid abusing my organs. I can recognize damage I've done to myself through sports injuries, secondhand smoke, too much sugar and caffeine, and so on. But as long as the damage is not too severe, I can apply a restorative ecological approach to correcting those ailments. I can make health decisions with the same rigor I apply to environmental stewardship, emphasizing the webs of interdependence involved in both.

Maintaining a forest is a much bigger operation than maintaining a lawn, and I have a lot less control over a forest than I do over my health. But the goal is the same: restoring the natural processes to as close to what they were before damage was done.

In the case of forests, this means before human intervention. If we use natural resources effectively, we won't have to waste so much time, energy, and money striving for some ideal state that is most likely unsustainable and ultimately unhealthy. To survive on this planet, we have to discover the important natural processes and their roles in the biosphere, figure out which of them are out of balance and might be contributing to unhealthy human lives, and restore them to a state in which they can sustain themselves for the long-term benefit of our species.

The practice of stewardship begins at home and in the field, but education is critical. Children need to think about ecosystems and their own health in similar ways. Preventative measures are far more favorable than those based on hindsight. And when restoration is necessary, even small acts can sometimes make a big difference. For example, I participate in a statewide program called the New York State Forest Stewardship Plan, which helps to preserve forests by educating landowners about the economic and ecological value of their timber and watersheds. It was through this program of education and affiliation that I learned about the eradication of vines from the forests of New York. So now every time I walk through the woods, I carry along my trusty industrial-strength pruning shears.

Attitudes toward nature are learned, not instinctive, and healthy habits rarely come about by good fortune. A willingness to learn about the natural things we often take for granted is a first step toward embracing the practices encouraged by restorative ecology. For that reason, I acknowledge the importance of conservation efforts that set aside land as forever wild for no reason other than preserving a place of wilderness for future visitors. In other words, even if no human ever intends to harvest species from a region, or use nature's products in any way, wilderness should still be set aside for contemplation and the joy of getting closer to the natural processes unperturbed by humans. In this respect, I find a spiritual element in restorative ecology. It gives me a sense of connection with other species and scales of time far outside human experience.

I can't say exactly why I appreciate hemlock trees so much. Maybe it's because of the lovely shade and protection they provide from the hot sun in summer and the sheltering comfort they offer in winter. Maybe it's because of the majestic repose of their gigantic silhouettes.[5] Whatever the reason, when I can get away from my hectic schedule of music tours and academics, I head for that familiar hemlock stand in the Finger Lakes region of New York State, and as soon as I get there, it feels as if I haven't a care in the world. I'm surrounded by towering mature timber, sheltered from wind, snow, or rain, among the giants of the eastern forests. Standing in a hemlock woodland gives me a sense of time travel, as if I'm stepping into a prehistorical landscape. Each individual tree has weathered hundreds of seasons, and hemlock species have existed much longer than has ours. In that sense, I feel like a visitor on my own farm.

The hemlock forests on my farm are a remnant of the Pleistocene era. Most of the large mammals in North America went extinct at the end of the Pleistocene. We are a large mammal, too, and how our species will fare is anyone's guess. There are as many signs pointing to our rapid extinction as there are to a hopeful future. Scientific research is teaching us what happened to the large mammals of North America, but time is probably running out for us. With the huge increase in our numbers and our impact on ecosystems, a new epoch of earth history is dawning.

+

Roughly eighteen thousand years ago, huge expanses of North America were cold and wet, perfect for hemlock woodlands. During this time, much of upstate New York, including my farm, was dominated by continental glaciers. Slabs of ice thousands of feet thick extended over the northern half of the continent. In the east, only the highest peaks of the northern Appalachians rose above the ice and snow.[6] In the Great Lakes region, an immense slab of ice weighed heavily on the earth. Even the Rocky Mountains and the Sierra Nevada in what is today California had glaciers thousands of feet thick.

Over the last two million years, successive periods of glaciation gripped the planet. Where the glaciers reached their farthest point south, called glacial maxima, they left piles of boulders, cobbles, and finer sands and soil known as terminal moraines. By mapping these moraines, geologists have delimited the extent of the glacial advances during the Pleistocene—the period from roughly two million years ago to the warming of the last few thousand years.[7] And by dating organic material in the moraines, they have generated reliable time frames for each advance and retreat.

Beginning about 23,000 years ago in North America, the rate of melting began to exceed the accumulation of snow, and within a few thousand years the ice sheets began a rapid northward retreat. Huge lakes and rivers formed from the snowmelt, and habitable land began to emerge around their highland banks and peripheries. As the ice sheets melted away toward the north, they left expansive, flat, outwash plains in their wake, full of thick, unconsolidated sediment that would later become fertile farmland.

The mountains reemerged from under the melting sheets of ice, and their valleys filled with lakes and rushing streams. Today, average global temperature is so warm that ice sheets occur only at high latitudes or very high altitudes. All other places are too mild for snow and ice to last through the spring, summer, and fall. Where immense glaciers once stood, now are continental expanses of terrestrial ecosystems full of plants and animals or huge freshwater wetlands and lakes. Today clouds spill their burden of water more often as rain on the land than as snow on accumulating glaciers.

Vestiges of the Pleistocene are still visible, if you know where to look. As the ice sheets retreated poleward, warmth-loving species expanded their ranges toward the north. Remnant populations of cold-loving species found safe havens on north-facing slopes, even at temperate latitudes like upstate New York or northern California. These relict species came into contact with the expanding ranges of temperate species, creating new webs of interaction that resulted in the ecosystems of our modern world.

Life today is a remnant of history. Since hemlocks occupy cold, north-facing slopes, we can hypothesize that they had a much greater range during times of extensive glaciation. This hypothesis has been supported by the discovery of their fossils in areas that today are too warm for glaciers or hemlock. Hemlocks know nothing of their ancestral conditions, but they will persist in their patchy grandeur until someday, perhaps, the cold returns again and they resume the inexorable expansion of their range.

Animals can be viewed through the same lens. Their ranges expanded and contracted in concert with the ebb and flow of climatic variation. For example, insects have great difficulty with ice and snow. With no ability to generate body heat or withstand the devastation of frozen tissues, they must stay away from habitats where winters are too harsh. Many larvae can make it through short cold snaps, due to their ability to metabolize and secrete an alcohol that acts as an antifreeze. But the greatest diversity of insects is found in warmer parts of the globe, particularly within tropical latitudes.

As global warming has occurred in recent decades, new combinations of species have come into contact, sometimes resulting in catastrophe. Numerous infestations and blights are affecting forests all over North America, often due to insect or fungus species encountering tree species that are accustomed to colder winter temperatures. In the past, extreme winters prevented the pests from causing problems. They couldn't establish a large enough population to damage the forests. But with milder temperatures, insects and fungi are causing significant damage, wiping out some species entirely and significantly limiting the geographic range of others. The mountain pine beetle is destroying lodgepole pines throughout the western United States. The emerald ash borer is devastating the ash trees of the eastern United States. The beech scale insect and its associated fungus have infested and wiped out virtually all of the mature specimens of the once-magnificent beech trees. As species make new contact with one another, their population structure and geographic ranges can change drastically.

When the population density of a species declines to a level that inhibits successful reproduction, extinction is not far off. Only fifty years ago, majestic American elm trees grew throughout the eastern hardwood forests of North America. Almost every community in the United States established in the 1800s has a main thoroughfare called "Elm Street" that was named after the huge trees lining their sidewalks. Because of a fungal blight, spread by an immigrant Asian bark beetle, not a single elm remains on those streets, and by 1970, virtually all of the mature American elms in the forests were gone as well.[8]

Whenever I escape to the sheltering hemlocks of our farm to contemplate the interactions of species past and present, I can't help but think of mastodons, those extinct elephants of the Pleistocene that only recently (geologically speaking) went extinct. They roamed throughout this area not so long ago, and many of them died in tranquil, marshy settings near lakes or rivers. Today their fossils turn up in bogs—mucky bottomlands surrounded by low hills, which collect organic remains like fallen leaves, perennial plants, and, occasionally, animals. Bogs generally have very slow rates of erosion, which means that they are excellent places for fossilization. Bogs are like organic graveyards that record thousands of years of dying organisms.

Near our farm, mastodons made big news not too long ago. In 1999, two different rural families were enlarging ponds on their property using excavating equipment. The logical place to locate ponds in this region are wet bogs. The most complete mastodon fossil ever discovered came from one of these pond-enlargement projects.[9]

Our farm has a pristine bog. Whatever fell in there over the past fifteen thousand years is likely still there, because the bog is drained only by the trickle of a tiny stream. If I ever wanted to bring in some huge excavating machines and destroy the fragile habitat, I probably could find the remnants of many large mammals. But my affinity for fossils doesn't go that far.

Mastodons and the members of the elephant family both descend from a common ancestor that lived more than 40 million years ago, probably in Africa. Today only two genera of elephants have survived: African elephants and Asian elephants. Neither group of living elephants has geographic ranges that come anywhere near the North American fossils of the massive beasts that roamed our farmland.

Mastodons lived in a variety of habitats, but they were common throughout North America during the Pleistocene. They occupied the coniferous forests of upstate New York, the grasslands and prairies of Wisconsin, river valleys in Texas, and coastal plains in Florida. During the Pleistocene, North America had a collection of large mammals even more impressive than that of modern-day Africa. Other species included the dire wolf (a huge version of the same genus that we know today as the northern timber wolf), the saber-toothed cat, giant sloth, bison, caribou, horse, and camel. One of the best places to see the Pleistocene megafauna fossils is in Los Angeles, at the La Brea Tar Pits. This spectacular museum showcases mounted skeletons and ongoing excavations that detail the interrelationships of these extinct animals. Instead of dying in bog deposits, the La Brea animals died in oil seeps. Areas around Los Angeles are pervaded by huge cracks in the earth's crust, which have been in place for millions of years. These zones of crustal weakness acted as conduits for buried oil to percolate up through sediment and spill out into ponds that dotted the landscape. The oil was thickly viscous, like tar, and if a mega-mammal fell into one of these ponds, it couldn't get out. Often a dying mega-mammal would attract all kinds of predators and scavengers, who would also get caught in the tar pool. Luckily, tar is an excellent way to preserve bones.

The fossils tell a grim story. Almost all of the large mammals that roamed North America suddenly went extinct at the end of the Pleistocene. Between about 15,000 and 13,000 years ago, for example, the mastodons experienced a rapid extinction over most of their species range. This was a time of global climatic upheaval that drastically affected all animals and plants. The glaciers had been retreating from their last glacial maximum. But even as the climate grew gradually warmer, severe cold snaps lasting 1,000 years or more caused glacial conditions to return temporarily. Within this sequence of freezing and thawing, the mega-mammals seemed to thrive, keeping pace with the shifting geographic ranges of the plant species they ate. Although they were probably adept at living in other habitats, there is no question that the bulk of mastodon localities was associated with hemlock and spruce forests. Evidence for this is the presence of these tree species in the stomach contents of their fossilized remains and the coexistence of white spruce and balsam fir fossils alongside their bones. Spruce species resemble the typical Christmas tree. They form part of the community of species in luxuriant forests called taiga found in cold northern latitudes. In the coldness that preceded the Recent epoch—also called the Holocene, starting roughly 11,600 years ago and leading up to today—spruce forests occurred much farther south than they do today. Most of the Great Plains were blanketed by this hardy, evergreen species. Taiga seems to have been the favored habitat of the mastodon. Today taiga is found only in subarctic regions of Canada and Russia or in high mountain ranges of the American West. During the Pleistocene, as the glaciers retreated toward the pole, the spruce forests followed suit, and the mastodon populations headed northward along with the shifting geographic ranges of the tree species.

The last cold period of the Pleistocene took place contemporaneously with the extinction of mastodons and numerous other large mammals between 12,900 and 11,600 years ago. This colder, dryer period in a generally warming climate was associated with large bodies of standing freshwater and wetlands similar to habitats we see today in Canada, Scandinavia, and Russia. Fossil plants reveal that a rapid transition occurred from tundra to spruce forest just before the onset of the cold snap, suggesting that a volatile climate typified the end of the Pleistocene.

The gradual warming that resumed about 11,600 years ago marked a major reorganization of life on the planet. The mastodon was one member of roughly thirty genera of mammals that went extinct at the beginning of the Holocene. Indeed, the Holocene marks one of the most significant mass extinction episodes in earth's history, because humans were affected by it. "Paleoindians" and other hunters around the globe subsisted on many of the large species that became extinct.[10]

During periods of glacial maxima in the Pleistocene, roughly 30 percent of the earth's surface was covered by thick sheets of ice. When so much water is frozen in continent-sized ice sheets, sea level drops, because the hydrological reservoir of our planet is finite. Areas where the sea today is shallow were exposed as dry land during glacial periods.

The depth of the Bering Strait between Alaska and Russia is relatively shallow. During the peak of the Pleistocene glaciations, the bottom of the sea in today's Bering Strait was above sea level. As a result, there existed a "land bridge" between the continents (though really this "bridge" was just a low plain). Migrating populations of widely roaming terrestrial animals crossed this land bridge, which is how mastodons were able to colonize North America after their diaspora from Africa and their colonization of Asia.

Another animal came across the Bering land bridge toward
the end of the Pleistocene: the brainy biped known as *Homo sapiens*. And the role of humans in the demise of the North American
megafauna constitutes one of the most enduring mystery stories
of our hemisphere. Could it be that we overhunted the large
mammals of the Pleistocene? If so, have we learned to curtail our
destructive ways when it comes to other species?

+

The time of the extinction of the mastodon corresponds to the cultural extinction of a civilization of North American humans. These
so-called Clovis people were nomadic and followed the megafauna,
often exploiting particular species for food and clothing. Hundreds
of archaeological sites scattered across North America contain Clovis arrowheads fashioned out of stone, flint, or volcanic glass. These
primitive artifacts do not occur at any locality later than the cold
snap just preceding the Holocene. The Clovis people seemed to
get along just fine during the relatively warm period leading up to
the cold snap. Thereafter, their way of life was finished, as was the
existence of many of the mega-mammals.

Numerous hypotheses have been advanced to account for the
coincident termination of the Clovis culture, the extinction of
the mega-mammals, and the Pleistocene-ending cold spell. Evidence of severe drought due to the warming trend just before the
deep chill suggests that the Clovis people and the mammals they
hunted were already under duress. Maybe they couldn't adjust
to the rapid drop in temperature. Another hypothesis is that humans overhunted the mammals, contributing to the extinction of
the mastodon and their own demise.[11]

Even though many large mammals went extinct, humans endured and switched to other ways of life. At least six cultural groups came after the Clovis people. The more recent cultures of the Holocene had to accommodate to a much reduced variety of large game animals. Their mammal of choice became the smaller American bison (*Bison bison*), a direct descendant of a subspecies of bison that persisted through the ecological catastrophe of the big chill at the end of the Pleistocene.

The American bison was well-suited to the gradual warming of the Great Plains grasslands of North America during the Holocene. North American Indians of the modern era, descendants of the Clovis Paleoindians of the Pleistocene, depended on this species for their way of life. When Europeans made contact with the Great Plains Indians, they saw roaming herds of American bison numbering in the millions. The population size of the species was estimated to be 60 million at the time of European contact with North America. By 1880, only a thousand American bison remained. Almost the entire species was wiped out through overzealous slaughter by nonnative Americans. Efforts to expand American civilization entailed military-style campaigns to eradicate the buffalo (another name for *Bison bison*). The tacit understanding was that Indians got in the way of American progress, so by removing the buffalo, Americans would remove the Indian way of life. The practice went hand in hand with the attempted ethnic cleansing of native peoples from eastern states, which also was sponsored by the U.S. military throughout the eighteenth century.

Thanks to conservation efforts beginning in the late nineteenth century, and the restoration ecology of more recent decades, numerous private herds of bison were kept and bred, and large areas, such as Yellowstone National Park, have been set aside as bison habitat. Today more than 350,000 bison live in North America. It is an impressive number, but it is less than 1 percent of the original population. Furthermore, the genetic diversity of today's herd is drastically reduced, and many bison today are hybrids with domestic cows carrying genes of distantly related European species.

Many large mammals have had a tough time sharing the earth with humans. Many are threatened or endangered, including the polar bear, grizzly bear, timber wolf, musk ox, and caribou. In this respect, we are still experiencing—and contributing to—the Pleistocene extinction.

+

When I'm in my favorite hemlock grove along the creek on our farm, I'm reminded of two things:

1. Species have interacted throughout history, compelled by climatic shifts and global events that entailed no act of grand or intelligent design. Species have no goal in mind. In this respect, the human populations of the past were no different than the animals they hunted. Humans have always acted on impulsive, shortsighted, immediate needs without recognizing the impact of their behavior on other species.

2. What we witness in nature is the consequence of those interactions. But today, given our ability to study the past more carefully than ever before using the methods and instruments of modern science, we can implement policies and practices to avoid some of the mistakes of our ancestors.

I have a tendency to apply these observations to things other than the natural world. When Brett and I talk about the popularity of Bad Religion, we sometimes use ecological analogies. We consider our audience a precious and finite resource, like a fishery. For example, the fisheries of Peru in the 1970s were among the world's most productive for anchovies, which are used in all kinds of animal feeds.[12] Because of overzealous fishermen from all nations, who flocked there to haul in unlimited amounts of fish, the anchovy population crashed and the fishery had to be closed. It took more than twenty years to rebuild the population of anchovies, during which time thousands of workers in the fishing industry were out of work and the markets for seafood changed dramatically. Only by respecting the balance of the ecosystem can we hope for a sustainable commercial fishing strategy in Peru.

Whenever we prepare to go on tour or produce a new Bad Religion record, we think about the negative prospects of overmilking our fans. We liken this overmilking to overhunting during the late Pleistocene or overfishing off the Peruvian coast. We respect our fans' intelligence and their desire to see and hear something new and special from us. Without our "core" fans, the band could not continue. We need to cultivate them by offering them new songs and playing live concerts for them with the hope that their enthusiasm for us will grow. Maybe they will tell their friends about us and our overall audience will grow. If we take them for granted and don't offer them our best effort, if we do shows without rehearsing or put out an album of half-baked songs, our fans might show up, but they probably will leave disappointed and never show up again. Our audience could vanish in a single album cycle. It's like the greedy carelessness of commercial fishing. Instead of cultivating a healthy relationship with their fans, some bands exploit their previous popularity and squeeze every last bit of loyalty from fans who grew tired of the "same old song" long ago. We approach our fans with the same respect I try to extend to the natural world. I am pretty sure they will turn out for our next concert if I remain committed to improving my skills and musical craft, just as I know those hemlocks will provide me with shade and solace so long as I continue to clip the parasitic vines away from their trunks.

I'm lucky enough to be able to spend a good amount of time on each side of the United States. Los Angeles, where I teach and record music, is not as tranquil as upstate New York, but I still make it a habit to bring my family and friends on an annual backpacking trip to the Sierra Nevada. We usually spend a week in the alpine zone, experiencing life alongside hardy plants and animals and studying the geology above the tree line. We also always find time to visit the magnificent stands of giant sequoia on the western slope of the Sierras. These trees make the eastern hemlocks seem like dwarfs. They are the largest living things on earth, rising nearly three hundred feet into the air. Many have trunks wider than a driveway, and they run straight up eighty feet before reaching the first branch!

Sequoia gigantea nearly went extinct a hundred years ago due to careless timber harvesters. Since the trees take nearly a thousand years to mature, it isn't possible to create a sustainable timber harvest of giant sequoia, and today the few remaining groves are protected by law. The groves are on north-facing slopes along canyons of the western Sierra Nevada between elevations of three thousand and eight thousand feet. Outside of this elevation, the seeds don't germinate, so no new stands will be able to take root except in the microhabitats where they are now located. Their habitat, like that of the eastern hemlock in the gorges of upstate New York, consists of pockets of cool temperatures, indirect sunlight, and wet forest floors. They represent the isolated remnants of a once-larger population that spanned a much broader geographic range throughout the Pleistocene.

Standing among those remnant populations, it is impossible not to conclude that we are somehow a part of all this. Some would call this a "spiritual connection"—the sense of being part of some larger web of life. Whatever you want to call it, the feeling is inescapable that we are living among the leftovers of a recent mass extinction. This realization is as emotionally moving to me as, I'm sure, the realization of God's will was to my great-grandpa Zerr.

All species make do with their current circumstances. Given grim surroundings and paltry education, can we hold anyone responsible if they choose not to live a life dedicated to restorative environmental practices and instead opt for short-term rapacity and greed? As a humanist, I tend not to hold people responsible for such a decision. All people have to make do with the social and economic tools given to them by their culture and their parents. If they aren't taught to think sustainably, why should I expect them to care about anything other than low prices and adequate pay?

As a musician, I've seen many bands strike it rich with a hit song. Then, when it comes time to make another album, they milk their hit song for all it's worth by penning similar songs instead of further developing their creative potential. Quickly the public loses interest and dubs the band a "one-hit wonder." Long-term creativity is annihilated in the interest of short-term capitalization and profit.

Perhaps we cannot hold individuals responsible for the short-term exploitation of our planet, but we certainly can change educational standards and orient young people toward a narrative of natural science. Humans have always been capable of drastic environmental perturbation and destruction to other species. Today our potential to exterminate other species is far more devastating than that hypothesized by the Clovis overhunting scenario. Whether we like it or not, we are stewards of the planet. We can't control all of the factors that promote healthy species. But we must recognize that nonhuman species need to interact without drastic interference from us for nature's inherent creativity to flourish.

As I was writing this, the governor of New York was considering provisions that would allow gas companies to drill and extract a massive buried reserve of natural gas that underlies most of New York State. Our farm is in a prime drilling location. The method of extraction, called hydrofracking, involves pumping a high-pressure cocktail of water and toxic chemicals thousands of feet below ground to fracture the rocks and release the natural gas. Vast areas of Colorado and Utah have been destroyed by this technique, which pollutes drinking water, creates huge pools of liquid waste many acres in size, and leads to other kinds of industrial pollution and deforestation. I understand the potential benefits of using natural gas: fewer greenhouse emissions than gasoline, cleaner delivery, and domestic supply as opposed to reliance on foreign countries. But the longer-term costs of extracting this nonrenewable resource from fragile areas of New York's forests are much greater than the gas companies admit. Polluted watersheds don't quickly rebound, and with agricultural needs increasing as population increases, the farmlands of New York

State should be left for food production and for biodiversity concerns instead of degraded for gas extraction.

Modern human societies have the ability to reflect on our own destructive practices. We don't have to suffer the same fate as the Clovis hunters. We can manage our resources better than they did. Modern humans understand more about earth processes and the careful balance of life on the planet than at any time in our history. The time will come—some say it has already passed, because of overpopulation and environmental change—when humans will need to actively manage all of our ecological resources. Agricultural, industrial, and governmental decision-making in the twenty-first century will inevitably require knowledge of the natural sciences. We need at least to consider the naturalist worldview and its implications as we debate whether we want a fact-based society or not. And all of us will need to be better educated about biology, which can't be a bad thing.

It is arrogant to think that we know how to manipulate complex ecosystems in an eternally sustainable way. We don't know, for instance, all the creative processes that occurred in the past and resulted in the current state of the world. But we know when we are being wantonly destructive, and there is no justification for it. We know when we are creating radical strains on populations of plants and animals for nothing but short-term profit for a few and misleading promises of wealth and comfort. We can wait until more environmentally friendly methods of natural-resource extraction are invented. We can recycle and reuse to a much larger extent. Instead of spending tens of millions of dollars on lawyers to block environmental preservation, companies can spend that money on research, fact-finding, and testing. The best public relations is found in the methods of the naturalist worldview: observation, experiment, and verification If citizens could verify that drilling is safe, for instance, there would be no environmentally based reason to oppose the gas companies.

By demonstrating in our laws and policies that nature is a process, not a thing to be exploited, we will say to future generations that we have learned something from the study of natural science and that we care about something greater than our own selfish needs. And then we will be remembered by later generations for our wisdom rather than our rapacity.

CHAPTER 9

A MEANINGFUL AFTERLIFE

Perhaps . . . a shift in values can be achieved by reappraising things unnatural, tame, and confined in terms of things natural, wild, and free.
    —Aldo Leopold, one month before he died in 1948[1]

[T]he intellectual workman forms his own self as he works toward the perfection of his craft. . . . [Y]ou must learn to use your life experience in your intellectual work: continually to examine and interpret it. In this sense craftsmanship is the center of yourself and you are personally involved in every intellectual product upon which you may work.
    —C. Wright Mills[2]

Many religions dwell on what happens to you when you die. In these religions—Hinduism, for example, with its belief in reincarnation, or traditional Christianity with its belief in heaven and hell—your actions on earth determine the disposition of your soul after death. Good actions are rewarded postmortem, and bad actions are punished. In essence, God is the parent in these religions and we are children. If we misbehave, we get an eternal time-out. This projection of typical child-parent interactions from our earthly experience to the supernatural realm seems to be why so many religious people believe that atheists are amoral hedonists. Without the check on behavior afforded by the afterlife, they say, people will act like spoiled infants.

The naturalist worldview sees no prospect of life after death, since no evidence supports the idea that people have immaterial, transcendental souls or that anything considered alive persists after death. The only thing that is transferred from one organism to another is information in the form of DNA and other biological molecules or cultural documents, artifacts, and traditions. I would like to think that my soul or spirit will live on in my music or writing, but that's not the kind of immortality religious people covet. The belief in souls by religious people appears to be another lifelong remnant of childhood misconceptions—in this case, that all living things have an essence that is separate from the physical body of that organism.

But just because naturalists do not believe in a life after death does not mean that they don't care what happens after they die. I am deeply concerned, for instance, about whether my family members will be happy and successful after I am gone, whether my friends will continue the traditions we have established, and whether the world will be a better place because of my actions. I hope that what I do in this life will make a long-term difference in the world, though I will never know whether this ambition is fulfilled. In fact, a strong case can be made that naturalists tend to care more about these things than do religious people, since naturalists are committed to an ethic that emphasizes the causal effects of our actions in the here and now, as opposed to a mythological hope for a better life in a supernatural realm. A core belief of naturalism is that this life is the only one we will ever experience, and therefore any hope for the betterment of our lives and the lives of others must come in this lifetime.

Many religious people say that focusing only on this life is not enough. Without the promise of heaven and the threat of hell, we don't have enough incentive to live a good life. I believe otherwise. Most of us may not have much significance in the causal events of the universe at large. (I am reminded of this each time I go backpacking in the wilderness.) But we have great significance to those close to us. Furthermore, we are more culturally connected than ever before in earth's history, which means that the chance any individual has to make a lasting difference in the world has never been greater. We all should devote as much effort as we can to fulfilling our potential as parents, as friends, and as members of the human species.

I have a friend—another naturalist—who thinks that potential fitness is the most important thing in life. This isn't the kind of fitness that makes your heart healthier. In conventional formulations of evolutionary biology, the number of viable offspring one creates is a mathematical measure of fitness. A really fit animal or plant will have lots of offspring that go on to produce more offspring in later generations.

My friend studies "herps"—reptiles and amphibians. He has concluded, from a lifetime of fieldwork, that the males with the highest fitness are the ones who have access to the most females. To him, human social interactions have one main rationale—which males can attract the most females. With lizards, the males who get the most females are those who hold and defend the largest and most bountiful territory. If a male lizard can create a boundary around a particularly luxuriant area and keep out all other males, females will be attracted to the resources in that area and he will have sole access to them for mating.

One reason my friend and I get along so well is that he is also a kick-ass guitar player who had a career in the decade before Bad Religion in a long-haired, Avocado Mafia–style rock band.[3] He has melded his experiences in music with his career as a naturalist and professor to arrive at a satisfyingly coherent picture of male-female relations: "The males with the best territory get the most chicks, and there is no better territory for humans than up on that stage." I have concurred but have added a caveat: "The front man, not the guitarist, has the best territory."

I have only two children, so my fitness, technically, is just average. But think for a minute about what it means to have children, even though not everyone will. If the human population is to stay the same size or grow, everyone has to have an average of at least two children. Of course, some people have no children and some people have more than two children. But let's assume for now that everyone has two children and that the differences average out over time. It's a pretty good assumption—the average fertility rate in the United States is currently about 2.05 children per woman.[4]

If you have two children and each of your children has two children, then you'll have four grandchildren. If each of your grandchildren has two children, then you'll have eight great-grandchildren. This doubling continues with each generation. You'll have sixteen great-great-grandchildren, thirty-two great-great-great-grandchildren, and so on. You can see this effect in obituaries—many people, if they live long enough, leave behind multiple grandchildren and great-grandchildren, and the numbers of their descendants will continue to increase after their deaths.

These numbers soon get very large. You won't live to see them all, but the numbers of your descendants can grow very quickly. Ten generations after a person dies—after somewhere between two and three centuries—that person will have more than a thousand descendants if each of the progeny has just two children. That doesn't seem right, but again there is plenty of evidence to back it up. If the average generation time is twenty-five years, then the *Mayflower* landed sixteen generations ago. Genealogists can't trace all of the descendants of people who arrived on the *Mayflower*—no records exist for the vast majority of humans who have lived during the history of our species. But genealogists know that each person on the *Mayflower* who has descendants living today has many thousands of descendants, just as would be predicted by the simple mathematics outlined above. Similarly, if you have children and your descendants have an average number of children, you will have many thousands of descendants a few hundred years from now.

Where will all of your descendants live? The most likely answer is that they'll live somewhere close to where you live today. In that case, your descendants will eventually start marrying one another, especially as they come to comprise a significant fraction of the total population in the area. They probably won't know that they are distant cousins (the kinship definition of "cousin" is two people who share a common ancestor, which, in this case, will be you), because they will have different last names and the records of their descent from you will be lost. But they should recognize that everyone who gets married is a cousin to his or her spouse, because all of us share common ancestors, whether in the more recent or more distant past.

Some of your descendants will not live where you live today. They will live in some other town, or some other state, or some other country, or some other continent. They will marry someone in the place where they live and start having children. In that case, you will begin to have multiple descendants in that town, or that state, or that country, or that continent. And within ten generations from that point, you will have more than a thousand descendants who live not where you live today but in that other part of the world.

Do you see where this is headed? Pretty soon the whole world starts filling up with your descendants. The population in the part of the world where you live today becomes more and more descended from you. And your descendants fan out into the rest of the world to start the same process elsewhere.

Until a few years ago, no one knew how quickly this process happened. But in 2004, a statistician, a computer scientist, and the coauthor of this book used some newly developed mathematical and computer-science techniques to model human ancestry. They were astounded by their conclusions. It turns out that a couple of thousand years from now, so long as you have children and they have children, everyone in the entire world will be descended from you.[5] This conclusion doesn't even require the assumption that everyone has two children—it happens no matter how many children people have. It requires only that people move around a bit over the course of their lives.[6] Given that assumption, it's a mathematical certainty that everyone in the world will be descended from you just a couple of millennia from now.

This observation also applies to people who lived 2,000 to 3,000 years ago. In other words, if someone lived in Athens at the time of Socrates (469–399 B.C.), or in China at the time of Confucius (551–471 B.C.), or even in southern Africa or South America 2,500 years ago, and had four or five grandchildren, then that person is the ancestor of everyone living on the planet today. So if Jesus did have children with Mary Magdalene, then he and she are the ancestors of most, if not all, of the people living on earth today—despite what Dan Brown's book *The Da Vinci Code* might say. Similarly, most, if not all, of the people in the world today are descended from Julius Caesar, Nefertiti, Emperor Gaozu of Han, and anyone else who lived and had children more than 2,000 or 3,000 years ago.

This seems counterintuitive, because we are taught to think of ourselves as the unique product of a direct line of ancestors. We tend to overstate the contribution of our famous relatives and forget the others. But, of course, there is no such thing as a direct line of descent. Think of all the countless combinations of reproductive events that took place leading up to your birth—not just your mom and your dad but your grandparents, your great-grandparents, and so on into the past. Typically, we think that we are descended from just a handful of the people who lived on earth two thousand or three thousand years ago, not from every person living then who has ancestors today. We are comfortable with the thought that our ancestors were a small subset of the human population. It's more common to hear someone say "I'm Irish" or "I'm of German and French descent" than to hear "I am a blend of everything, a mongrel, a mutt." That's because most of us identify with the relatively recent cultural traditions of a particular group. But the biological data tell a different story. We have unbelievably complicated genealogies. We are part of a vast web of ancestry that is far more interconnected than we have ever imagined.

✦

Our social networks are also far denser and more interconnected than we imagine, regardless of whether we have children or not.

These networks form a web of meaning analogous to the webs of ancestry in which we are embedded. But this meaning takes shape contemporaneously, not over multiple generations. As such, our social connections and the influence of our behavior on others are subject to change on a completely different timescale than our genealogical relations. In most cases, they also are more important than our family history in the complex set of causes that determine who we are.

How many people do you think you have meaningful social interactions with on a regular basis? I mean people with whom you exchange an opinion, information, or even just social niceties. All of them are affected in one way or another by your interchanges, whether the effects are large or small. Certainly this list would include family members, friends, coworkers, and even casual acquaintances. Let's be extremely conservative and say that ten people fall into this list, though the actual number is almost certainly much higher.

Each of the people on your list also has at least ten (and probably far more) people with whom he or she interacts. Some of these people will be the same as the people on your list, but some will be different. So your "second-order" circle of influence—the people you influence plus the people they influence—is already between ten and ninety people and probably much higher.

Many of these people live in your community, but at least some will live far away. These distant connections might be a family member who lives in another town, a "friend" on Facebook from a different country, a business associate in another state, or a college roommate who moved far away. These individuals become social seeds who can spread your influence in other parts of the world.

As with networks of ancestry, these social connections saturate the world with amazing rapidity. Instead of occurring over generations, as in ancestral relations, your social influence can effect a wide-ranging response within days or even hours. The Internet has made this influence especially obvious. When people had to rely on telegraphs, letters, word of mouth, and other narrow-bandwidth means of communication, a war in a small African country, a disease outbreak in a remote part of China, or a comparably momentous event could go completely unreported outside the borders of the affected country. Contrast that with today's wide-bandwidth style of information transfer. News, shards of information, and the most trivial minutiae bombard us constantly no matter where we turn. The smallest happenings in remote places, such as a YouTube video that features a stray dog interrupting a soccer match in Paraguay or video from a mining accident in an Austrian village, can be disseminated to people all over the world. Small stories shared on the Internet or on cell phones can, if they go viral, reach a worldwide audience.

Even in the absence of the modern methods of communication, studies have shown that you are just a handful of links away from any other person anywhere in the world. This observation is the basis of the "six degrees of separation" concept—the idea that any two people can be linked through a chain of no more than six acquaintances.[7] In other words, it's possible to say, "X knows someone who knows someone who knows someone who knows someone who knows someone who knows Y," and the terms "X" and "Y" can be replaced with the names of any two people anywhere in the world and the statement will be true. Actually, research has added some important qualifications to this statement.[8] If two people are in different social classes or speak different languages, the number of links can be greater. Also, finding the shortest link between two people is a highly nontrivial problem, as mathematicians say, and can require immense computer power to solve.[9]

But the density of our social networks is undoubtedly a fact, as several new sources of data have shown. For example, records of calls made on cell phones render social networks visible and amply demonstrate the tightly linked interconnections of large numbers of people.[10] And many people who use Facebook or some other social networking technology intuitively understand that they could reach anyone else on the network in a very small number of steps.

These networks of human connections are not just an interesting feature of our lives. They *are* our lives. I recognize that the people reading this book who communicate on Twitter may never come in contact with a youngster in a remote Amazon Basin *barracca*. But no matter your circumstances, rich or poor, urban or rural, developing nation or G7, your social network plays the key role in forming your worldview and in creating your sense of right and wrong. And today the likelihood that people of vastly different demographics might meet, at least electronically, is higher than at any time in the past, due to the interconnectivity of mass media.

None of us could exist outside this web of social connection, no matter how much we may tout our individuality. We start interacting with others the moment we are born—with our mothers and fathers, most immediately, and then with grandparents, siblings, peers, teachers. Our interactions with others grow as we start watching television, reading books, and listening to music. As adults, we enter into professions, develop romantic ties, and hone our worldviews. We are immersed in a social context that gives meaning and direction to our lives. And because of our intense social connections, we can't help but have an influence on future generations, no matter what we do with our lives.

It's my firm conclusion that human meaning comes from humans, not from a supernatural source. After we die, our hopes for an afterlife reside in the social networks that we influenced while we were alive. If we influence people in a positive way—even if our social web is only as big as a nuclear family—others will want to emulate us and pass on our ideas, manners, or lifestyle to future generations. This is more than enough motivation for me to do good things in my life and teach my children to do the same.

One way to describe our lives is to say that meaning arises from the complicated social web in which we are immersed. But this description seems entirely inadequate. Science, religion, ethics, art—all are products of the ways that humans interact. Furthermore, these social institutions change with each successive round of creative combinations, and there is far more potential for social change today than at any time in the past, due to the vastly increased number of social connections created by modern technology. What many believe to be rock-solid social institutions must be viewed as plastic, subject to deformation, and, as with science, liable to restructuring. As Aldo Leopold stated in the epigraph to this chapter, there is good reason to believe that a shift in values will be achieved as the collection of verifiable knowledge proceeds. All institutions must recognize this basic fact of modern life.

For religious believers, this can be a dispiriting view. If my great-grandfather Zerr were alive today, I would ask him, "How can you build a faith based on the supposed immutability of sacred text in a society where scientific knowledge contradicts that text and where the social fabric is changing before our eyes?" According to my mom, late in his life, Grandpa Zerr's insistence on Bible literalism created distance between him and many in his flock who desired a more progressive, tolerant style of biblical interpretation. As with many orthodox theists, his approach ended up alienating many people from the social web he was trying to create. In today's world, it is increasingly difficult for any institution—religions, universities, newspapers, manufacturers, or even music genres like punk rock—to cling too stubbornly to absolutist principles of what "should be." The changes that dictate "what is"—including the creative combinations that come from vastly enhanced social networks—are occurring with blinding speed. It may seem impossible to keep up with all the changes, but I find it is difficult only if you insist on a static worldview. Once you embrace the dynamism inherent in an increasingly connected world, the changes are not as worrisome. And this dynamic social view fits perfectly with the naturalist principles of observation, experimentation, and verification.

There is a tension between our membership in dense social networks and our autonomy as individuals. Social expectations can be stifling. Throughout this book, I've been calling attention to the ways in which individuality can be suppressed by the demands of others. How can we resist becoming mere cogs in the grand web of social mechanisms that underlie society?

My answer is simple. We need to embrace our creativity and stop trying to fight it. When unexpected things arise, whether representing tragedy or opportunity, we have to exercise wise decision-making toward creative ends. Each of us has the potential to make an impact on the world through our actions and words. Whether you're a parent raising a child, a minister delivering a sermon, a bricklayer building a wall, a biologist exploring nature, an author writing a book, or a professor teaching a class, there have never been more opportunities for humans to share their ideas.

I feel particularly fortunate in this regard, because I became a musician and songwriter and have been able to express many of my ideas in songs. What is a song, after all? It's a distillation of experience, emotion, and worldview shared with an audience. A popular song is one that affects people in such a way that they can internalize it and adopt the song as a meaningful part of themselves. A song might have a temporary effect, or it may become a lifelong companion. "That's my song" is the ultimate compliment for any songwriter who has composed something that others enjoy so much they have adopted it as their own.

Songwriting is a craft, and it should come as no surprise that those who dedicate themselves to it improve over time. If you were building a house and wanted to construct a beautiful handmade stairway, would you hire someone under thirty? A famous woodworker's saying goes, "The lyf so short, the craft so long to lerne."[11] Older carpenters and stonemasons almost always do a better job than younger craftspeople, because they view each job as an experience from which to learn. The next job is an opportunity to improve on what they did in their last job. The same can be said of serious songwriting. I've always believed that my best work is waiting just around the corner. Even if I'm successful in writing a pleasing melody and good lyrics, I remain motivated to use satisfying elements from past songs in new and different ways.

I have had great good fortune with Bad Religion throughout

the thirty years of the band's existence. In 2001, Brett rejoined the band, contributing to a new burst of creativity that has led to what I think are some of our best albums. The demand for us to perform has never been greater, and we regularly travel throughout the United States and around the world to do shows. I've been friends with Brett, Jay, Greg Hetson, Brian, and the newest member of the band, the immensely talented drummer Brooks Wackerman, for decades. We are like a family. Together we have done some remarkable things.

On many fortunate occasions, I've heard people singing along with one of my songs on their car stereos with their windows rolled down, unaware that I was in the car next to them. At those moments, I felt a great sense of satisfaction. I knew that I had created, even if just for a moment, something meaningful for another person's life. Somehow my worldview, or, at the very least, a melody I'd written, was translated by them in evocative ways. There's no telling how deeply they understood what I was trying to say, or if they internalized any of the meaning of the lyrics, but somehow the song had struck a nerve. Maybe their life was enhanced by my music. Maybe they shared my words with their friends or loved ones. Maybe they built on my creativity in adding to a wider social web that will persist long after I am gone. In my opinion, there is no greater hope for an afterlife than being remembered by the people you touched, the things you did, and the ideas you shared. You don't have to be a singer or even a public figure to enjoy such an afterlife. You only have to enhance the relationships you already have. By doing so, you can be confident that you will become part of something bigger than yourself. And after you die, people will remember you and talk about you and extend your influence to future generations.

Creativity is a challenge. It requires us to be fully human—autonomous yet engaged, independent yet interdependent. Creativity bridges the conflict between our individuality and our sociality. It celebrates the commonality of our species while simultaneously setting us apart as unique individuals. We exercise our ability to change social networks while strengthening and reinforcing those networks.

Life is an act of endless creativity. With all its simmering tragedy and occasional catastrophe, a human life is an amazing thing to contemplate and experience. None of us had any special plan laid out for us when we were born. By abandoning the idea that an intelligent designer created us, we can wake with each dawn and say, "What's done is done. Now how can I make the best of the here and now?" Life is never static. Despite catastrophic tragedies, life has persisted in evolving new varieties of unimaginable forms. I find comfort in the narrative of evolutionary history. When I create, I feel that I am a participant in the grand pageant of life, a part of the ongoing creative engine of the universe. I don't know if that feeling is enough to replace the solace of religion in the lives of most people, but it is for me.

ACKNOWLEDGMENTS

I used to think statements in the preface that read "All the ideas, mistakes, and opinions in here are my own" were patently obvious to the reader, so why include it? Now, after completing this book, I empathize with all the authors who wrote those disclaimers. I can see that they were simply trying to protect the reputations of the teachers from whom they received their education. This book is a distillation of all the interactions I've had with many people over the years, and I can't adequately express my profound appreciation for their insights and wisdom—even though I might have interpreted it in ways they didn't expect. I offer these acknowledgments as an extension of my gratitude to those who have influenced my thinking and helped me bring this project together.

This project could not have gotten off the ground had it not been for the persistence of my agent, Marc Gerald of the Agency Group, who phoned and e-mailed me for eighteen months suggesting a lunch meeting "when you're ready" to discuss my book idea. After that meeting, Caroline Greeven, also of the Agency Group, went to work crafting an initial outline from hundreds of pages of my past writing. I give so many warm thanks to Marc and Caroline for believing in me and in the project we discussed at that Barney Greengrass lunch meeting.

When we pitched the idea to Bob Miller, Julia Cheiffetz, and

Debbie Stier at HarperStudio, I wasn't aware that my project was about to be embraced by a company of such consummate professionals and visionary leadership. After getting to know them better, I can honestly say that there is no other publisher with whom I would rather work. Thanks to Bob, Julia, Debbie, and Jessica for guiding the early phase of this project.

When I decided to bring Steve Olson on board as coauthor, I knew I was ready to take this book "to the next level." Steve's award-winning writing and his knack for scientific exposition was something I admired. We met each other many years ago at a Bad Religion concert. We talked about human ancestry—the topic of his best-known book—and later he melded his knowledge of evolution and punk rock by writing two articles about me that were published by *Wired* and *Paste* magazines. As we began the task of putting together my ideas for *Anarchy Evolution,* Steve had some worries that came out only after the project was complete. He had never coauthored a book, but he had read a lot about other collaborators who had nightmarish experiences with similar projects. The ease with which *Anarchy Evolution* came together, however, shocked both of us. There were no snags, no hold-ups, no arguments. Each chapter went through several iterations without a single hitch. Thanks a lot, Steve, for being such a great coauthor and friend.

Working with Carrie Kania at It Books has been a pleasure. Her vision has helped this project immeasurably. The publicity was masterfully handled by Greg Kubie at HarperCollins and Austin Griswold at Epitaph Records. Kevin Callahan did a fantastic job overseeing the marketing, and Katie Salisbury handled the logistics with aplomb.

The manuscript was improved greatly by comments from the

following people who read through the entire first draft. I thank Megan Shull, Will Provine, Paul Abramson, Jay Phelan, Preston Jones, and Lynn Olson. I am also grateful for the fantastic skill and attention of my editor Julia Cheiffetz, who read numerous revisions of the book and expertly navigated it. Olga Gardner Galvin did a great job of copyediting the manuscript.

My episodic academic life has been enhanced by numerous people, but I thank particularly Jay Phelan, Mark Gold, Will Provine, Warren Allmon, Fritz Hertel, Peter Vaughn, Laurie Vitt, and Paul Abramson for being that rare combination of academic colleagues and great friends. To the UCLA administrators in the "front office"—Tracy Newman, Lily Yanez, and Lauri Holbrook—thanks for making teaching so enjoyable. To the students whom I felt privileged to teach in the Life Sciences Core Curriculum and in Earth and Space Sciences at UCLA: thank you for studying and especially for coming to my office hours to discuss philosophical issues that weren't on your tests.

The process of songwriting is a special craft, and I've been particularly fortunate to have a cowriter who is also a childhood friend and who functions more like a brother than a business partner: thanks, Brett Gurewitz, for all your encouragement and sagacity. I've benefited greatly from our constant engagement in brisk debate and lively discourse on philosophical and musical topics.

For being those rarest of gems, true friends who also participate in the great creative enterprise of touring musicians, I thank my bandmates and crew: Jay Bentley, Greg Hetson, Brian Baker, Brooks Wackerman, Jens Geiger, Cathy Mason, and Ron Kimball. In all the countless hours we have spent together on five continents, I've always enjoyed our lighthearted ramblings and unity when it mattered most. In other areas of my professional career, I thank Steven Barlevi, Eric Greenspan, Frank Nuti, and Darryl Eaton for being good friends and important advisers.

No project can be considered a success without the loving support of a family. The encouragement I get from my family is unending, and I thank them for allowing my work to intrude on so many aspects of our relationship. Graham and Ella, a lot of this book was written with you in mind. Even if you never read it, know that its content was inspired in a big way by my love for you. Alli, I love and appreciate you more than the words in this book can express. Thank you for all your patience and understanding. Mom and Dad and Grant, you inspired me to know more, do more, and create more, and in the process I discovered the depths of your wisdom. Thanks also go to my extended "family," those friends of the "inner circle" who are my constant support group, my most trusted companions, and mirth-makers of the highest caliber: Wrye Martin, David Bragger, and Megan Shull. I also send thanks with great appreciation to my other family, Frank and Sheila Kleinheinz, for their love and support.

NOTES

Chapter 1: The Problem with Authority

1. Laplace stated this in response to Emperor Napoleon I's question about why he had not mentioned God in the astronomical analysis of orbital motion in his five-volume *Treatise on Celestial Mechanics* (1825). This statement has been widely quoted, especially in contemporary literature. The quotation seems to have become more relevant as the importance of the Enlightenment became better understood. See Brian L. Silver, *The Ascent of Science* (New York: Oxford University Press, 1998), 61.

2. Einstein made this remark in 1930 in conversation with a friend. See A. Calaprice, ed., *The Expanded Quotable Einstein* (Princeton, New Jersey: Princeton University Press, 2000), 14.

3. In science, authoritative positions come from paradigms, the currently held views of the field. In essence, paradigms are dogma. Scientists who defend paradigms too stringently often give the impression, contrary to the spirit of science, that dogmatic authority should be respected. In fact, science depends on paradigm shifts that come about through challenge to dogma and an undermining of previous authority. Eventually, through discovery, verification, and novel theorizing, standard scientific practices become worn out, and new paradigms emerge during extraordinary episodes called "scientific revolutions." See Thomas S. Kuhn, *The Structure of Scientific Revolutions* (Chicago: University of Chicago Press, 1962).

4. Phil Zuckerman, "Atheism: Contemporary Numbers and Pat-

terns," in Michael Martin, ed., *The Cambridge Companion to Atheism* (New York: Cambridge University Press, 2007), 47–65.

5. Plato makes this argument in the dialogue *Euthyphro,* in which Socrates finds flaws in various definitions of piety. "Euthyphro . . . is the caricature of a popular 'pietist' who knows exactly what the gods wish. To Socrates' question 'What is piety and what is impiety?' he is made to answer: 'Piety is acting as I do!' " From Karl Popper, *The Open Society and Its Enemies, Volume I. The Spell of Plato* (London: Routledge, 1945), 265. The dilemma of Euthyphro can be paraphrased by asking, "Does something deserve special status because it is favored by God, or is it favored by God because it has special status?" This was Socrates' dilemma with respect to piety, as related by Plato.

6. Physicists commonly combine space and time into a single entity called the space-time continuum. In this respect, there are only three things in the universe: space-time, matter, and energy.

7. For more on the unpredictability of complex systems, see Stephen Wolfram, *A New Kind of Science* (Champaign, Illinois: Wolfram Media, 2002).

8. Herbert Vetter, *Speak Out Against the New Right* (Cambridge, Massachusetts: Harvard Square Library, 2004).

9. Jacquetta Hawkes, *The Atlas of Early Man* (New York: St. Martin's Press, 1976).

10. Richard Leakey and Roger Lewin, *Origins* (New York: Dutton, 1977).

11. Donald Johanson and James Shreeve, *Lucy's Child: The Discovery of a Human Ancestor* (New York: Viking, 1989).

12. In 1902, Rudyard Kipling published a collection of stories that described various whimsical or magical ways in which animals acquired certain characteristics. For example, the rhinoceros got bumpy skin and a grumpy disposition because a Parsee filled his skin with cake crumbs when the rhinoceros took it off to go swimming. Evolutionary biologists use the term "just-so story"

to refer to causal explanations of evolutionary outcomes that are impossible to verify or falsify.

13. As far as we humans can tell, our species is the only one that can ponder its own existence and recognize, isolate, and experiment with many elements of the physical and biological world. In that sense, and in the sense that we are an exceedingly active species that has populated the globe and can explore outer space, humans are justified in their self-congratulation as one of evolution's greatest products. But there certainly are other criteria for determining evolutionary "success" (a discussion that, in my opinion, is best done over cocktails or at the after-party). For instance, under the heading of "resilience," our species has not been tested adequately. Our genus, *Homo,* has only been around for roughly 2 million years, not a very long time by most standards of measuring generic age. Consider that some genera, such as *Lingula*—a burrowing, marine, shelled invertebrate—has been busy churning up the nearshore sediment for nearly 600 million years, and you get the idea that other species can make us look like infants who haven't even begun to experience the kinds of ecological upheavals that the earth has to offer.

Also, our metabolism is pathetically unidimensional, in some ways. We must consume oxygen by breathing, and we must eat other things to garner enough nutrients. If humans are deprived of oxygen, biochemical changes in blood carbon dioxide and acidity begin to affect brain function within minutes. Death of all cells in the body is not far off unless oxygen flow is promptly restored to the lungs.

Consider that the first organisms to appear on earth, more than 3.4 billion years ago, are still with us today. These microscopic organisms, called cyanobacteria, make all their own necessary biochemical "building blocks" through photosynthesis. They don't have to eat for their source of carbon. Rather, they consume carbon dioxide from the atmosphere and use the energy from the sun to split water molecules. Unlike humans, cyanobacteria see oxygen as simply a waste product! They need water, however, in the same way that humans need oxygen (as a source of electrons for energy metabolism). However, if cyanobacteria are denied their favorite electron source—water—they can switch to another chemical, hydrogen sulfide, or they can use molecular hydrogen as an electron donor. In other words, they have metabolic abilities that allow them to withstand drastic shifts in the chemical makeup of their environment. Cyanobacteria are equally comfortable in aerobic environments (oxygen-rich) or anaerobic environments (oxygen-lacking), brightly sunlit or dark (they can reduce elemental sulfur in the dark). This remarkable metabolic flexibility is far beyond our species' capacity, and it's probably why cyanobacteria have existed during such a tremendous span of earth's history. Given the high metabolic needs of humans and our limited tolerance to environmental changes, we would have very little inherent flexibility if atmospheric composition were to undergo dramatic shifts.

14. Theodosius Dobzhansky, "Nothing in Biology Makes Sense Except in the Light of Evolution," *The American Biology Teacher* 35 (March 1973), 125–129.

15. "Deep evolutionary time" is a term used to distinguish historical periods of millions to hundreds of millions of years. Evolutionary changes also occur in "ecological time," tens to hundreds of thousands of years. These are sometimes referred to as "microevolutionary" changes. But the patterns in the fossil record, to which I'm referring in this passage, are macroevolutionary, large-scale changes that affect entire evolutionary lineages.

Chapter 2: Making Sense of Life

1. From an interview in Cambridge, Massachusetts, on June 25, 2003, in Gregory W. Graffin, *Evolution, Monism, Atheism, and the Naturalist World-View* (Ithaca, New York: Polypterus Press, 2004), also published as Greg Graffin, *Evolution and Religion: Questioning the Beliefs of the World's Eminent Evolutionists* (Ithaca, New York: Polypterus Press, 2010). See www.polypterus.org.

2. From an interview in Bedford, Massachusetts, on June 25, 2003, ibid., 167.

3. Lynn Margulis and Dorion Sagan, *Acquiring Genomes: A Theory of the Origins of Species* (New York: Basic Books, 2002).

4. Good reviews of punk and its ascendancy can be found in Brian Cogan, *The Encyclopedia of Punk* (New York: Sterling Publishing, 2008), and *Mojo Magazine, Punk: The Whole Story* (London: Dorling Kindersley, 2006).

5. A recent review of eukaryotic evolution is by T. Martin Embley and William Martin, "Eukaryotic Evolution, Changes and Challenges," *Nature* 440 (2006), 623–630.

6. This statement will cause apoplexy in many evolutionary biologists, and understandably so. It comes dangerously close to the ultimate sin in evolutionary thinking, according to most leaders in the field. That transgression is called "teleology," an ancient holdover from the time of Aristotle. It boils down to a type of mysticism, having no basis in fact, which finds explanations of all things in relation to their purpose. An example might be that the purpose of grass is for cows to have a place to pasture and feed. The purpose of cows is to provide milk and meat for humans. The purpose of evolution, from a teleological perspective, is to create greater complexity of forms eventually then leading to human beings. This kind of purposive drive in evolution has never been demonstrated. Interested readers can consult Henri Bergson, *Creative Evolution,* Arthur Mitchell, trans. (New York: Henry Holt & Co., 1913), and Pierre Teilhard de Chardin, *The Phenomenon of Man,* Bernard Wall, trans. (New York: Harper & Row, 1959). For a discussion of problems with teleology, see Ernst Mayr, *The Growth of Biological Thought* (Cambridge, Massachusetts; London, England: Belknap, 1982), 528. In my opinion, all discussions of "adaptive design" and "optimality" in evolution perilously skirt the edges of teleology.

7. Edward B. Daeschler, Neal H. Shubin, and Farish A. Jenkins, "A Devonian Tetrapod-like Fish and the Origin of the Tetrapod Body Plan," *Nature* 440 (2006), 757–763.

8. Steve Olson, *Mapping Human History: Discovering the Past Through Our Genes* (Boston: Houghton Mifflin, 2002).

9. Graffin, *Evolution, Monism, Atheism, and the Naturalist World-View.*

10. Will wrote these comments to me when reviewing an early draft of this book.

11. See chapter 8, note 3, for more on the naturalistic fallacy.

12. I put quotation marks around the word "genes" because it has become difficult to define precisely what a gene is. It can be defined as a double-stranded, helicoidal nucleic acid molecule that contains the genetic code, but this is overly simplistic. See Mark B. Gerstein, Can Bruce, Joel S. Rozowsky, Deyou Zheng, Jiang Du, Jan O. Korbel, Olof Emanuelsson, Zhengdong D. Zhang, Sherman Weissman, and Michael Snyder, "What Is a Gene Post-ENCODE? History and updated definition," *Genome Research* 17 (2007), 669–681. It's also important to remember that another type of nucleic acid, RNA, uses the genetic code and contains genes. In humans, as in all eukaryotic organisms (those whose cells are eukaryotic—the outcome of the mutualistic symbiosis referred to in this chapter, as opposed to prokaryotic—those cells that do not undergo endosymbiosis), RNA serves crucial regulatory functions that help determine DNA expression in the synthesis of proteins.

   In general, we should think of genes as coded biochemical information that produces some kind of functional biological product. But there are many other potential sources of traits. See Mary Jane West-Eberhard, *Developmental Plasticity and Evolution* (Oxford: Oxford University Press, 2003), 20.

13. For more on the complexities of interactionist explanations, see Richard Lewontin, *The Triple Helix, Gene, Organism, and Environment* (Cambridge, Massachusetts: Harvard University Press, 2000), 116, and David S. Moore, *The Dependent Gene: The Fallacy of "Nature vs. Nurture"* (New York: W. H. Freeman, 2001).

14. Peter P. Vaughn, vertebrate paleontologist extraordinaire. My

master's thesis was also overseen and signed by the legendary Everett C. Olson as well as earth scientists Walter (Ted) Reed and Gerhard Oertel, all of UCLA. The title of my thesis was "A New Locality of Fossiliferous Harding Sandstone: Insights into the Earliest Vertebrate Environment and Some Aspects of Dermal Skeletal Tissue," University of California, Los Angeles, 1990.

15. Greg Graffin, "A New Locality of Fossiliferous Harding Sandstone: Evidence for Freshwater Ordovician Vertebrates," *Journal of Vertebrate Paleontology* 12 (1992), 1–10.

16. Neil Shubin, *Your Inner Fish: A Journey into the 3.5-Billion-Year History of the Human Body* (New York: Random House, 2008).

17. Adaptive radiation: "[T]he success of [an evolutionary] phyletic lineage to establish itself in numerous different niches and adaptive zones." Ernst Mayr, *What Evolution Is* (New York: Basic Books, 2001), 208.

18. Recent research on microorganisms has demonstrated that there are major problems in applying the "tree of life hypothesis" to all organisms. The tree of life (TOL) assumes that the relationships of all species can be traced in a stepwise, ancestor-descendant fashion and that there is a genetic signature that allows us to do so (genotype). The analogy with a family tree is easy to understand, but in reality the analogy leads one astray. A patriarch might have ten children who have ten children each. All hundred cousins would then share some DNA with the patriarch of the family as well as with one another. But it can't be conclusively demonstrated that all species alive today, especially if bacteria are included, have clear-cut ancestor-descendant relationships. Bacteria can exchange genetic material. Hence two different bacterial species can give rise to a third species in one generation. Because of this, some researchers suggest that life be viewed more as a "web of life" instead of the TOL. The TOL applies to the vast majority of familiar species, but even this might end up being insufficient as more data are collected from the biosphere. For a Web-based version of the tree, see http://tolweb.org/tree.

19. Darwin's and Wallace's theories were presented at a meeting of the Linnaean Society in 1858.

20. Cambridge University Press has been republishing some of the Bridgewater Treatises. See, for example, Bell, *The Hand: Its Mechanism and Vital Endowments as Evincing Design*.

21. Neal H. Shubin, Edward B. Daeschler, and Farish A. Jenkins, "The Pectoral Fin of *Tiktaalik roseae* and the Origin of the Tetrapod Limb," *Nature* 440 (2006), 747–749.

22. Moncure Daniel Conway, *Autobiography, Memories and Experiences of Moncure Daniel Conway, Vol. 1* (New York: Houghton Mifflin, 1904), 359.

23. For more background on James Leuba, see Gregory W. Graffin and William B. Provine, "Evolution, Religion, and Free Will," *American Scientist* 95 (July–August 2007), 294–297. Also see James H. Leuba, *The Belief in God and Immortality: A Psychological, Anthropological and Statistical Study* (Boston: Sherman, French and Co., 1916).

24. James H. Leuba, "Religious Beliefs of American Scientists," *Harper's Magazine* 169 (1934), 291–300.

25. I was interested in belief among the most highly respected authorities in the field of evolutionary biology, rather than all biologists or all scientists, which is why I restricted my study to members of the national academies around the world.

26. Edward J. Larson and Larry Witham, "Leading Scientists Still Reject God," *Nature* 394 (1998), 313.

Chapter 3: The False Idol of Natural Selection

1. William B. Provine, *The Origins of Theoretical Population Genetics* (Chicago: University of Chicago Press, 2001), 199 (afterword).

2. A recent popular introduction to taxonomy is by Carol Kaesuk Yoon, *Naming Nature: The Clash Between Instinct and Science* (New York: W. W. Norton, 2009).

3. I like to distinguish wisdom from knowledge, although I didn't

know how to do it as a high school student. The distinction is based on my reading of Bertrand Russell, *The Scientific Outlook* (New York: W. W. Norton, 1931). Wisdom helps me mold my physical and social surroundings into something I consider beneficial, whereas types of knowledge that do not help me do this I consider trivia.

4. Martin J. S. Rudwick, *The Meaning of Fossils,* 2nd ed. (Chicago: University of Chicago Press, 1985).

5. In his autobiography, Darwin wrote: "I can indeed hardly see how anyone ought to wish Christianity to be true; for if so the plain language of the text seems to show that the men who do not believe, and this would include my Father, Brother and almost all my best friends, will be everlastingly punished. . . . I will here give the vague conclusions to which I have driven. The old argument from design in nature, as given by Paley, which formerly seemed to me so conclusive, fails, now that the law of natural selection has been discovered. . . . There seems to be no more design in the variability of organic beings and in the action of natural selection, than in the course which the wind blows. Everything in nature is the result of fixed laws." Nora Barlow, ed., *Charles Darwin's Autobiography* (New York: W. W. Norton, 1958), 87. The person to whom he refers is William Paley, one of the leading natural theologians of the late eighteenth and early nineteenth centuries, who argued that a watch could not have been constructed without the intelligent handiwork of a watchmaker. He used this as a metaphor, with God as the handyman, to explain the intricate "contrivances"—what modern naturalists might call adaptations—in nature. For a recent criticism, see John O. Reiss, *Not by Design: Retiring Darwin's Watchmaker* (Los Angeles: University of California Press, 2009).

6. For brief accounts of the lives of Darwin's children, see http://www.aboutdarwin.com/darwin/Children.html.

7. Darwin surmised that all "animals are descended from at most only four or five progenitors, and plants from an equal or lesser

number. . . . [O]n the principle of natural selection with divergence of character, it does not seem incredible that, from such a low and intermediate form [an algae], both animals and plants may have been developed; and, if we admit this, we must likewise admit that all the organic beings which have ever lived on this earth may be descended from some one primordial form." Charles Darwin, *On the Origin of Species by Means of Natural Selection, or the Preservation of Favoured Races in the Struggle for Life*, 6th ed. (London: John Murray, 1884), 425.

8. In addition to countless conversations with departmental faculty, colleagues, and students stretching back in time throughout my graduate school days, my views have been and continue to be stimulated by numerous volumes dealing with the problem of natural selection. This is not "light" reading. I sometimes spend months reading single volumes, digesting small amounts each night before dinner (or on tours, reading in two-hour blocks during plane flights). For those interested in the more technical literature, see Eva Jablonka and Marion Lamb, *Evolution in Four Dimensions, Genetic, Epigenetic, Behavioral, and Symbolic Variation in the History of Life* (Cambridge, Massachusetts: MIT Press, 2005); Mary Jane West-Eberhard, *Developmental Plasticity and Evolution* (Oxford: Oxford University Press, 2003); John A. Endler, *Natural Selection in the Wild* (Princeton, New Jersey: Princeton University Press, 1986); and Reiss, *Not by Design: Retiring Darwin's Watchmaker*.

9. Most often, even though evolutionists rarely explicitly state it, new evidence is incorporated into a narrative that sheds light on evolutionary theory. The interpretation is based on an "actualistic model." This term is one of the components of an underlying principle of evolution called "uniformitarianism," which has two factors: (1) earth processes in the past were dictated by the same natural laws that operate today (actualism), and (2) those processes occur at the same pace and intensity as they do today (gradualism). Most scientists reject the second element of uni-

formitarianism, because evidence doesn't support it. But almost all discoveries are consistent with actualism as an underlying principle. In evolution, for instance, we can use an actualistic model—by studying the growth stages of juvenile animals alive today in order to interpret a fossil that is millions of years old—to deduce that our fossil is the jaw of a juvenile mammal, even though we have no other evidence of the fossil individual except its jaw and a single un-erupted molar. See Donald R. Prothero and Fred Schwab, *Sedimentary Geology, An Introduction to Sedimentary Rocks and Stratigraphy*, 2nd ed. (New York: W. H. Freeman and Co., 2004), 454.

10. Barbara Forrest and Paul R. Gross, *Creationism's Trojan Horse: The Wedge of Intelligent Design* (New York: Oxford University Press, 2004).

11. John Angus Campbell and Stephen C. Meyer, *Darwinism, Design, and Public Education* (East Lansing: Michigan State University Press, 2003).

12. Ernst Mayr and William B. Provine, eds., *The Evolutionary Synthesis: Perspectives on the Unification of Biology* (Cambridge, Massachusetts: Harvard University Press, 1980).

13. For the development of the idea of "fitness surfaces" and its inventor, Sewall Wright, see William B. Provine, *Sewall Wright and Evolutionary Biology* (Chicago: University of Chicago Press, 1986).

14. See R. C. Lewontin, *The Genetic Basis of Evolutionary Change* (New York: Columbia University Press, 1974).

15. Nina G. Jablonski and George Chaplin, "The Evolution of Human Skin Coloration," *Journal of Human Evolution* 39 (2000), 57–106.

16. Kenichi Aoki, "Sexual Selection as a Cause of Human Skin Colour Variation: Darwin's Hypothesis Revisited," *Annals of Human Biology* 29 (2002), 589–608.

17. Timothy D. Weaver, Charles C. Roseman, and Chris B. Stringer, "Were Neanderthal and Modern Human Cranial Differences Produced by Natural Selection or Genetic Drift?" *Journal of Human Evolution* 53 (2007), 135–145.

18. For information about methylation and nongenetic inheritance, see Jablonka and Lamb, *Evolution in Four Dimensions*.

19. M. J. West-Eberhard, *Developmental Plasticity and Evolution* (New York: Oxford University Press, 2003).

20. F. J. Odling-Smee, K. N. Lalaud, and M. W. Feldman, *Niche Construction: The Neglected Process in Evolution* (Princeton, New Jersey: Princeton University Press, 2003).

21. At the time, Epitaph was a label in name only, one that we all came up with as a formality to brand our first two records. Epitaph has since grown into one of the most important independent record labels in the world thanks to the work of its owner, my friend and fellow songwriter Brett, who dropped out of high school, took the equivalency exam, and went on to become an artist, recording engineer, and record mogul.

22. An article in *Spin* magazine gave some insight into the Dogtown skateboarders and their adoption of punk music in the early 1980s:

> In the mid-'70s, [Jay] Adams and [Tony] Alva were a step ahead of everyone, pioneers of skating's hard-core approach to life in general. But in the late '70s, the punks caught up. "Black Flag, Circle Jerks, Descendents, Bad Religion, Suicidal Tendencies. We picked up on all the music that was happening in L.A. at that time," says Alva. "There was so much energy at those shows. Skate and punk fed off each other because they were both total outlets for aggression."
>
> Punk replaced Ted Nugent and Jimi Hendrix as the soundtrack for skate sessions; the music paralleled the sessions themselves, which had been turning more and more violent as the Dogtowners' hard-core reputations preceded them. "A lot of people were gunning for us, because they'd read about us in the magazines," Alva says. "We were like a rolling ball of chaos, this mobile gang on a recon mission. We'd show up at a skatepark somewhere, and there'd be guys who'd come up to us and get in our faces, telling us we weren't so hard-core."

Naturally, a fight would ensue. At night, after going to shows for local bands like Suicidal Tendencies (whose lead singer was Jim Muir's younger brother Mike), things got even more violent. "We'd go to parties, take Quaaludes, get in fights with bats and stuff," says Adams. . . .

In the most unfortunate incident to punctuate that era, Adams's good luck finally failed to coincide with his bad behavior. By 1982, he had developed a taste for tequila and ruining other people's nights. One evening, a trashed Adams and some punk friends found a pair of gay men walking down the street to yell at. When the men yelled back, Adams started kicking one while a friend punched the other. In a few moments, both pedestrians lay facedown on the concrete. Others at the scene soon joined in, kicking the two prone men with their steel-toed boots. By the time they were finished, one of the men was dead. Two days after the incident, Adams was arrested at his apartment and charged with murder, though he insisted he had left the scene by the time the others started kicking the men. He was ultimately convicted of assault, for which he served four months in jail. (From G. Beato, "The Lords of Dogtown," *Spin* 15, no. 3 [March 1999], 114–121.)

Chapter 4: The False Idol of Atheism

1. The actual translation is from the original French: *L'Homme n'est malheureux que parce qu'il meconnoit la Nature. Son Esprit est tellement infecte de prejuges qu'on le croiroit pour toujours condamne a l'erreur: le bandeau de l'opinion, dont on le couvre des l'enfance lui est si fortement attache, que c'est avec la plus grande difficulte qu'on peut le lui oter.* From Paul Henri Thiery, Baron d' Holbach, *Systeme de la Nature, ou Des Loix du Monde Physique & du Monde Moral; par M. Mirabaud, nouvelle edition* (Londres, preface, 1771). An English translation is available from Kessinger Publishing online.

2. Sam Harris, *The End of Faith: Religion, Terror, and the Future of Reason* (New York: W. W. Norton, 2004). Richard Dawkins, *The God Delusion* (Boston: Houghton Mifflin, 2006). Daniel C. Dennett, *Breaking the Spell: Religion as a Natural Phenomenon* (New York: Viking, 2006). Christopher Hitchens *God Is Not Great: How Religion Poisons Everything* (New York: Twelve, 2007). A good example of the kinds of conversations that take place among atheists is a video called "The Four Horsemen," available at http://www.RichardDawkins.net, which features Harris, Dawkins, Dennett, and Hitchens.

3. Penny Edgell, Joseph Gerteis, and Douglas Hartmann, "Atheists as 'Others': Moral Boundaries and Cultural Membership in American Society," *American Sociological Review* 71 (2006), 211–234.

4. Barry A. Kosmin and Ariela Keysar, *American Religious Identification Survey 2008* (Hartford, Connecticut: Trinity College, 2009).

5. Graffin, *Evolution, Monism, Atheism, and the Naturalist World-View,* 120–121.

6. Dawkins introduced the word "meme" in his book *The Selfish Gene.* A meme is, like a gene, transmitted from one person to another, but not through sperm and egg. Rather, a meme is transmitted in the form of cultural symbols, words, or behaviors. Think of it as "an idea that sticks," for some reason or other. Memes have nothing to do with transmission of genes. Dawkins, however, tacitly hypothesizes in this passage that there might be a gene or a group of genes that affects one's brain in a way that makes it more susceptible to social directives or reverence to authority. "Mimetic exploitation" refers to a hypothetical situation whereby an idea or directive is readily taken up by a person. Hence that person has the "susceptible-to-gullibility gene." Again, this is all hypothetical.

    Curiously, "meme" is found in the *Oxford English Dictionary* and is defined there as a term from biology. It is not, however, found in the most recent biology textbooks we use to teach undergraduates at UCLA: David Sadava et al., *Life the Science of Biology*, 8th ed. (Sunderland, Massachusetts: Sinauer and Associates, 2008), and Jay Phelan, *What Is Life, A Guide to Biology* (New York: W. H. Freeman, 2009). For the original reference, see Richard Dawkins, *The Selfish Gene* (Oxford: Oxford University Press, 1976), 192.

7. It is possible that the brain contains certain "cognitive structures" that are innate or present from birth. If this is true—and this is the central claim of the cognitive psychologists—then the social "imprinting" that occurs during the first few years of life would provide only a relatively subtle effect on one's worldview later in life, since presumably the innate structure of one's worldview would have been established at birth. If, however, deeply held beliefs are the result of personal life experiences—and are relatively uninfluenced by innate brain structures—then the early social environment and education could play a key role in determining one's worldview. For more reading on this and other brain-related topics, see Jean-Pierre Changeux, *Neuronal Man* (New York: Pantheon, 1985); Francis Crick, *The Astonishing Hypothesis* (New York: Scribner, 1994); Antonio R. Damasio, *Descartes' Error: Emotion, Reason, and the Human Brain* (New York: Avon, 1994); Gerald M. Edelman and Giulio Tononi, *A Universe of Consciousness, How Matter Becomes Imagination* (New York: Basic Books, 2000); Joseph LeDoux, *Synaptic Self, How Our Brains Become Who We Are* (New York: Viking, 2002); and Steven Pinker, *The Blank Slate* (New York: Penguin, 2003).

8. Vassilis Saroglou and Antonio Muñoz-García, "Individual Differences in Religion and Spirituality: An Issue of Personality Traits and/or Values," *Journal for the Scientific Study of Religion* 47 (2008), 83–101. See also Bruce Hunsberger, Michael Pratt, and S. Mark Pancer, "A Longitudinal Study of Religious Doubts in High School and Beyond: Relationships, Stability, and Searching for Answers," *Journal for the Scientific Study of Religion* 41 (2002), 255–266.

9. Bob Altemeyer and Bruce E. Hunsberger, *Amazing Conversions: Why Some Turn to Faith and Others Abandon Religion* (Amherst, New York: Prometheus Books, 1997). Also see Altemeyer and Hunsberger, *Atheists: A Groundbreaking Study of America's Nonbelievers* (Amherst, New York: Prometheus Books, 1997).

10. Frank Newport, "This Christmas, 78% of Americans Identify as Christian," Gallup, 2009.

11. Brian J. Grim and David Masci, "The Demographics of Faith," Pew Research Center, Washington, D.C., 2008.

12. The numbers in this paragraph and the following one are from Phil Zuckerman, "Atheism: Contemporary Numbers and Patterns," in *The Cambridge Companion to Atheism*, Michael Martin, ed. (New York: Cambridge University Press, 2007), 47–65.

13. There are many good scientific reasons that justify my belief that music has deep significance in the way people think. See Anthony Storr, *Music and the Mind* (New York: Free Press, 1992), and Oliver Sacks, *Musicophilia*: *Tales of Music and the Brain* (New York: Alfred A. Knopf, 2007).

14. The song "Come Join Us" from the Bad Religion album *The Gray Race* is a sarcastic rant about going with the crowd:
    so you say you gotta know why the world goes 'round
    and you can't find the truth in the things you've found
    and you're scared shitless 'cause evil abounds
    come join us
    well I heard you were looking for a place to fit in
    full of adherent people with the same objective
    a family to cling to and call brethren
    come and join us
    all we want to do is change your mind
    all you need to do is close your eyes
    come join us, come join us, come join us
    don't you see all the trouble that most people are in
    and that they just want you for their own advantage

but I swear to you we're different from all of them
come join us
I can tell you are lookin' for a way to live
where truth is determined by consensus
full of codified arbitrary directives
come join us
all we want to have is your small mind
turn it into one of our kind
you can go through life adrift and alone,
desperate, desolate, on your own
but we're lookin' for a few more stalwart clones
come join us, come join us, come join us
we've got spite and dedication as a vehement brew,
the world hates us, well we hate them too
but you're exempted of course if you come join us
independent, self-contented, revolutionary, intellectual,
brave, strong and scholarly
if you're not one of them, you're us already so
come join us, come join us, come join us, come join us!

15. "Moral terrorism" comes from Hitchens's book, in the chapter titled "Is Religion Child Abuse?" See Hitchens, *God Is Not Great: How Religion Poisons Everything.* 218.

Chapter 5: Tragedy: The Construction of a Worldview

1. Robert J. Richards, *The Tragic Sense of Life, Ernst Haeckel and the Struggle over Evolutionary Thought* (Chicago: University of Chicago Press, 2008), 107.

2. This quote comes from a memorial written by Darwin only one week after his daughter's death. It can be found online at the Darwin Correspondence Project (http://www.darwinproject. ac.uk/death-of-anne-darwin#memorial), or in Sydney Smith and Frederick Burkhardt, eds., *The Correspondence of Charles Darwin,* vol. 4 (Cambridge, England: Cambridge University Press, 1989), Appendix II.

3. The focal point of historian Robert Richards's thesis is that Ernst Haeckel's influence on evolutionary theory—from the late nineteenth century all the way up to the current atheistic attitudes among modern evolutionary biologists—resulted from his central tragedy in life, the death of his beloved wife, Anna, from "pleurisy" (possibly a ruptured appendix). Richards acknowledges, as I do in this book, that atheism is not a foregone conclusion from the study of evolution. But he believes that Ernst Haeckel, as the most widely respected popularizer of Darwin's theory (he lived until 1913, almost twenty-five years longer than Thomas Henry Huxley, "Darwin's bulldog"), was responsible for the atheistic tone of modern evolutionary theory—despite his romantic idealism and monistic belief that "God and nature are one." Richards writes: "[My] thesis considers certain non-essential aspects of modern evolutionary theory, namely its materialistic and anti-religious features. These, I believe, are contingent cultural traits of the modern theory (of evolution). . . . [M]any of the early proponents of Darwinian theory were both spiritualist—that is, they accepted a non-materialistic metaphysics—and believers—that is, they integrated their scientific views with a definite, or sometimes indefinite, theology. Asa Gray, William James, and Conwy Lloyd Morgan are just a few prominent advocates of evolutionary theory who nevertheless rejected a stony, desiccated materialism." Ernst Haeckel, because of his tragedy, dismissed orthodox religion as superstition and advanced a militant monistic philosophy—full of romantic idealism, such as archetypes and creative nature. "My thesis is even more specific, namely: had Haeckel not suffered the tragic events . . . his own version of Darwinian theory would have lost its markedly hostile features and these features would not have bled over to the face turned toward the public." See Richards, *The Tragic Sense of Life, Ernst Haeckel and the Struggle over Evolutionary Thought,* 15–16.

4. The commentaries are still available from Guardian of Truth Foundation, Bowling Green, Kentucky (http://www.truthbooks.net).

5. Sometime during the nineteenth century, the Churches of Christ splintered into numerous sects in the United States. E. M. Zerr's church, for instance, belonged to the "non-instrumentalists" sect. The "instrumentalists" believed that it was okay to accompany the singing voice with instrumental music. In my mom's small town of Anderson, Indiana, there were other Churches of Christ, separated by only a few miles, that didn't communicate with one another. Part of the reason for this was that they disagreed on their views about music and singing. Other differences surrounded the function of ministers. E. M. Zerr's church had no ministers. Even though the church had officers, called elders, who were the most highly respected members, anyone could lead a service. The process was called "mutual edification." Services were held every Sunday morning, Sunday afternoon, and Wednesday night. Women could not be elders, but they were allowed to lead services.

Another difference between E. M. Zerr's sect and other Churches of Christ pertained to the involvement of Bible colleges. Some congregations were considered "institutional," because they encouraged their youngsters to attend a sanctioned Bible college, such as Bob Jones University. My great-grandfather's sect, however, had no such intentions. He believed that Bible study was an individual pursuit and didn't require the involvement of a Bible college. Still, he obviously supported education, since his daughter (my grandmother) and granddaughter (my mother) both became teachers and were educated at universities.

6. I recorded some of those songs on an album in 2006, titled *Cold as the Clay*, released on ANTI, a division of Epitaph Records.

7. See Thomas Lewis, Fari Amini, and Richard Lannon, *A General Theory of Love* (New York: Random House, 2000). The authors describe, in scientific terms, the biological causes for the feeling of love and its importance in shaping our emotional well-being.

8. See Jared Diamond, *Collapse, How Societies Choose to Fail or Succeed* (New York: Viking, 2005). One of his objectives in writ-

ing this book was the assertion that learning from history allows us to avoid future tragedies: "Globalization makes it impossible for modern societies to collapse in isolation, as did Easter Island and the Greenland Norse in the past. Any society in turmoil today, no matter how remote—think of Somalia or Afghanistan as examples—can cause trouble for prosperous societies on other continents, and is also subject to their influence. . . . For the first time in history, we face the risk of global decline. But we also are the first to enjoy the opportunity of learning quickly from developments in societies anywhere else in the world today, and from what has unfolded in societies at any time in the past" (23).

9. A similar observation made in the context of species rather than individuals appears in David M. Raup, "The Role of Extinction in Evolution," *Proceedings of the National Academy of Sciences, USA* 91 (2002), 6758–6763.

10. Donald R. Griffin, *Animal Minds* (Chicago: University of Chicago Press, 1992).

11. G. Brent Dalrymple, *Ancient Earth, Ancient Skies: The Age of the Earth and Its Cosmic Surroundings* (Stanford, California: Stanford University Press, 2004).

12. J. William Schopf and Bonnie M. Packer, "Early Archean (3.3-billion to 3.5-billion-year-old) Microfossils from Warrawoona Group, Australia," *Science* 237 (1987), 70–73. See also Martin D. Brasier et al., "Questioning the Evidence for Earth's Oldest Fossils," *Nature* 416 (2002), 76–81, and J. William Schopf et al., "Laser Raman Imagery of Earth's Earliest Fossils," *Nature* 416 (2002), 73–76.

13. The geologist was John Phillips (1800–1874), professor (reader) at the University of Oxford and president of the Geological Society of London from 1856 to 1860. See John Phillips, *Life on the Earth: Its Origin and Succession* (Cambridge, England: Macmillan and Co., 1860).

14. The effect of mass extinctions on single-celled organisms is not clearly understood. More data need to be collected.

15. See Luis W. Alvarez, "Experimental Evidence That an Asteroid Impact Led to the Extinction of Many Species 65 Million Years Ago," *Proceedings of the National Academy of Sciences, USA* 80 (1983), 627–642.

16. Michael Benton, *When Life Nearly Died: The Greatest Mass Extinction of All Time* (New York: Thames & Hudson, 2003).

17. David M. Raup, *Extinction: Bad Genes or Bad Luck?* (New York: W. W. Norton, 1991).

18. See Stephen Jay Gould, *Wonderful Life: The Burgess Shale and the Nature of History* (New York: W. W. Norton, 1989). This book explains the unpredictability of humans evolving from invertebrates, even though we can retrace the steps in the fossil record through which that evolution occurred. It demonstrated to me that making sense of where we came from doesn't imply anything about where we are headed.

19. One of my favorite essays on this topic is Will Provine, "No Free Will," *Isis: Catching Up with the Vision: Essays on the Occasion of the 75th Anniversary of the Founding of the History of Science Society* 90 (1999), S117–S132.

Chapter 6: Creativity, Not Creation

1. See Julian Huxley, *Evolution, the Modern Synthesis* (London: George Allen & Unwin, 1945), 458.

2. See Richard Dawkins, *The Blind Watchmaker, Why the Evidence of Evolution Reveals a Universe Without Design* (New York: W. W. Norton, 1996), 9.

3. I would like to draw the distinction between creativity—something that emerges from purposeless and unintended combinations of previous works—and utility—the struggle to innovate with the purpose of solving a problem. A lot of human endeavors that we think of as "the creative arts and sciences" are a combination of both creativity and utility. But the aspects of human life that I analogize with nature are the blind (with respect to purpose) creative acts that arise from whimsical experimentation

rather than from obsessive striving toward the solution to a particular task or problem. A lot of so-called hit songs emerge from randomness and whimsy as opposed to utilitarian concerns.

Todd Rundgren once told me that his song "Bang on the Drum All Day" came to him one morning upon his waking from a dream. "I don't even know why I recorded it that day," he said. It went on to become his best-known song and today is heard in commercials and at sporting events around the world. Likewise, Ric Ocasek related a story to me while we were discussing whether or not to include a song I wrote ("Punk Rock Song") on the album he was producing for Bad Religion. He stated that he had a similar dilemma with one of his songs ("Shake It Up") during the recording of one of the Cars albums. "I sure am glad we decided to include it," he said of his song, because it became the title track and biggest pop-hit song of their double-platinum album. We decided, on a whim, to include "Punk Rock Song," which became the album's "punk hit," and we still perform it today. None of these musical success stories was predicted. They were motivated by whimsy, and were not the intended highlights of the works.

Sometimes instrumentation, too, can come from unexpected experimentation. Our first album contains a song called "Fuck Armageddon, This Is Hell," with a slow-motion breakdown in the bridge section that features an acoustic piano. I wrote the song on a piano, so our producer, Jim Mankey, said, "Let's put some punk piano on the recording." None of our contemporaries had ever put acoustic piano on a hard-core punk album, and if we were in a less experimental mood that day, it might never have happened. But I played it, and the acoustic piano was featured on the song's final version. The song became a punk standard and was admired for its creative use of piano.

Many scientific discoveries also have emerged from the realm of creativity, as distinct from utility. The famous discoveries of penicillin, X-rays, and rubber, to name just a few, are well-known examples of unintended fortuity. It is true that some scientific and technological breakthroughs have come from years of research and testing, but in honest contemplation, we have to acknowledge the stark reminder that some of our most perplexing challenges—the common cold, flu, cancers, depression, and renewable energy, among many others—are still far from solutions, even after years of problem-solving activity. I predict that the cures and solutions will come eventually, and probably from experienced natural scientists. But those "aha!" moments will come from creative rather than utilitarian approaches.

To cite an evolutionary analogy, we can consider the small-scale, genetic shifts that occur from one generation to the next under the heading of "utilitarian changes" (i.e., microevolutionary). These changes are small, and may not lead to the formation of new species. The large-scale changes that are responsible for anatomical or physiological novelties—and the formation of new species and even higher taxonomic categories—can be considered under the heading of "creative changes" (i.e., macroevolutionary). Most of evolution is utilitarian, but the great diversity of life is due to creativity. This can be substantiated theoretically if one accepts even a rough version of Stephen Jay Gould and Niles Eldredge's theory of "punctuated equilibrium," which states that most evolutionary change occurs in short bursts of time (short bursts of innovative creativity), but the history of any evolutionary lineage is predominantly comprised of long periods of stasis (utilitarian). See Stephen Jay Gould, *The Structure of Evolutionary Theory* (Cambridge, Massachusetts: Belknap Press, 2002), chapter 9.

Precisely how these creative "revolutions"—also called "saltational" episodes—occur in evolution is still somewhat mysterious. But progress is being made. For example, it has been postulated, based on genetic and developmental discoveries in larval invertebrates, that fertile sexual encounters between animals of vastly different species have occurred in the past. The integration of one genome into that of another species (resulting in a "heterogenome") is a phenomenon that is raising awareness among biologists. Heterogenome formation among the freely dispersed sex cells of marine invertebrates might transpire every 10 million years or so, which means that at least fifty times in the history of life, a creative episode has happened. This figure accords well with fossil data and the diversity data of extant phyla. See Lynn Margulis and Dorion Sagan, *Acquiring Genomes: A Theory of the Origins of Species* (New York: Basic Books, 2002), chapter 10.

Another example in biological experimentation between species comes from America's most remarkable horticulturalist, Luther Burbank. See Jane S. Smith, *The Garden of Invention: Luther Burbank and the Business of Breeding Plants* (New York: Penguin Press, 2009).

In 1893, Burbank published a catalog titled *New Creations in Fruits and Flowers,* and it contained pictures and descriptions of plants that had never been seen before. He worked tirelessly for twenty-five years, first on his Massachusetts farm and later on his homestead in Santa Rosa, north of San Francisco, to develop new forms of plant life. These were not merely new varieties or races. By all practical methods of identification, what Burbank presented in that catalog were new species. See Luther Burbank, *Luther Burbank: His Methods and Discoveries and Their Practical Significance,* vol. 12 (New York: Luther Burbank Press, 1915), 128–134.

Debate still rages among biologists over the definitions of race, form, variety, subspecies, and species, not only among plants but among all living things. When scientists force one species to breed with another species (which is called a crossing), the resulting offspring is known as a hybrid. A lot of hybrids are sterile, which means that they cannot reproduce, usually because of incompatibilities between sperm and eggs of the same hybrid generation. In evolutionary terms, therefore, hybrids are dead ends. In nature, pollen from one individual might get distributed by wind onto numerous different species of flowers. Perhaps a hybrid will develop, but if that hybrid cannot produce offspring, then it is essentially a wasted generation from an evolutionary perspective.

Burbank's creativity was manifested in the combinations of forced breeding experiments that he undertook. He repeatedly introduced pollen from one species onto the female organs of other species. According to his records, just one plant in ten thousand might produce a viable offspring. But this tireless trial and error paid off. Burbank introduced scores of *viable* hybrids (able to reproduce).

When the public saw all of the plants that resulted from Burbank's hybrid crosses, the origins of new species became much more plausible. Burbank's catalog helped to convince a skeptical public about the authenticity of Darwin's evolution theory.

Today, some scientists still have a hard time accepting hybrids as new species because it contradicts the most commonly recited definition of species. For example, chimpanzees and humans share almost all the same genes but are considered different species. One reason is because they are "reproductively isolated" from each other. That is to say, members of one species cannot produce viable offspring with members of the other species. A viable hybrid organism violates this understanding of species. As Luther Burbank demonstrated, there is much about life we fail to see if we restrict ourselves too stubbornly to the reproductive-isolation definition of a species.

In fact, hybrids are plentiful in horticulture, in animal husbandry, and in nature. Hybrid speciation occurs with grizzly and polar bears, white-tailed and mule deer, the butterflies known as *Heliconius,* numerous finches and woodpeckers, and many freshwater and saltwater fish species. Agricultural examples are even more plentiful, including most prominently hybrid corn. What Luther Burbank began has become a precise industry of breeding and amplifying the most commercially viable hybrid plants and creating markets that depend on them.

The presence of well-documented hybrid species reveals that creativity is the hallmark of nature. Given the palette of reproductive potential, an almost unlimited variety of possible life-forms might be produced. I appreciate the unpredictability of all this. Think of the millions of fortuitous meetings between species going on at this very second. The potential for creativity is obviously immense.

4. John Emsley, *Nature's Building Blocks: An A–Z Guide to the Elements* (Oxford: Oxford University Press, 2001), 183.

5. Here I feel compelled to add a caveat from one of my favorite theories that borders on sci-fi. There is a curious characteristic of certain bacteria and viruses: they can withstand high doses of cosmic radiation and very low pressure nearing a perfect vacuum, the kind of conditions experienced in outer space. Svante Arrhenius advanced an idea in the early twentieth century that bacteria and viruses could be transported in interstellar space by radiation pressure from the sun's rays. This forms the basis for what is known as the "panspermia" theory. The famous astronomer Sir Fred Hoyle and biologist Chandra Wickramasinghe took up this idea and expanded it by reasoning that the earth is perfectly situated in our solar system to receive microbes from space that are shed from comets. Originally, such microbes might have rained down onto the early earth and found suitable conditions for replication, starting the process of organic evolution (other planets, such as Mars, had less favorable conditions for replication but possibly received just as much "seeding" from comets as earth did). At the core of Hoyle and Wickramasinghe's reasoning is the question: How could bacteria evolve "adaptations" to the conditions of outer space, when none of those conditions (i.e., X-rays and very low vacuum pressures) exists on earth, unless they, in fact, are traits that evolved to withstand the conditions of their origin, outer space? See Sir Fred Hoyle and Chandra Wickramasinghe, *Evolution from Space: A Theory of Cosmic Creationism* (New York: Simon and Schuster, 1981).

6. See Robert M. Hazen, *Genesis: The Scientific Quest for Life's Origins* (Washington, D.C.: Joseph Henry Press, 2006), or Andrew H. Knoll, *Life on a Young Planet, the First Three Billion Years of Evolution on Earth* (Princeton, New Jersey: Princeton University Press, 2003).

7. One way in which life could have originated is discussed in greater detail in Alonso Ricardo and Jack W. Szostak, "The Origin of Life on Earth," *Scientific American* 301 (September 2009), 54–61.

8. Today, anyone with Internet access can retrace our route on Google maps, Google Earth, or Bing maps. Search for "Riberalta, Bolivia," and follow the Madre de Dios westward.

Chapter 7: Where Faith Belongs

1. From the song "Rock Love" off the album *Adventures in Utopia* by Utopia, composed by Roger Powell, Todd Rundgren, Kasim Sulton, and John Wilcox (Bearsville Records, 1980).

2. From Bertrand Russell, *The Autobiography of Bertrand Russell: 1872–1914* (Boston: Atlantic Monthly Press, 1967), 3.

3. In the movie *This Is Spinal Tap*, directed by Rob Reiner, guitarist Nigel Tufnel is working out a piece on the piano, which he describes as a combination of Mozart and Bach—"like a Mach piece, really"—written in the key of B minor, the "saddest of all keys." He likes this key because it "makes people weep instantly." When asked the title of the piece, he says it's called "Lick My Love Pump."

4. See Ted Honderich, ed., *The Oxford Companion to Philosophy* (New York and Oxford: Oxford University Press, 1995), 838, for a discussion of "solipsism"—the view that only oneself exists.

5. Craig T. Palmer, "Mummers and Moshers: Two Rituals of Trust in Changing Social Environments," *Ethnology* 44 (2005), 147–166.

6. Celia A. Brownell and Claire B. Kopp, eds., *Socioemotional Development in the Toddler Years: Transitions and Transformations* (New York: Guilford, 2007).

7. Stephanie D. Preston and Frans B. M. de Wall, "The Communication of Emotions and the Possibility of Empathy in Animals," in Stephen G. Post, Lynn G. Underwood, Jeffrey P. Schloss, and William B. Hurlbut, eds., *Altruism and Altruistic Love: Science, Philosophy, and Religion in Dialogue* (New York: Oxford University Press, 2002).

8. Brett described a Ramones concert, which we attended at the Palladium as teenagers, in an article by Steve Appleford, "Live Nation's Crown Jewel: Hollywood Palladium Reopens This Week," *L.A. Weekly*, October 16, 2008.

9. Steven Weinberg, *The First Three Minutes: A Modern View of the Origin of the Universe* (New York: Basic Books, 1977).

10. Richard Dawkins, *River Out of Eden: A Darwinian View of Life* (New York: Basic Books, 1995).

11. See Steven Weinberg, *Dreams of a Final Theory: The Scientist's Search for the Ultimate Laws of Nature* (New York: Pantheon, 1992), and Richard Dawkins, *Unweaving the Rainbow: Science, Delusion, and the Appetite for Wonder* (Boston: Houghton Mifflin, 1998), and William B. Provine, "No Free Will."

Chapter 8: Believe Wisely

1. From an interview in Falmer, England, June 13, 2003. See Graffin, *Evolution, Monism, Atheism, and the Naturalist World-View*, 157.

2. From an interview in Bedford, Massachusetts, June 25, 2003. See ibid., 167.

3. An inevitable philosophical discussion, briefly mentioned in chapter 2, must be dealt with here. It concerns the objection to my insinuation that "nature is good." Philosophers like to discuss something called the "naturalistic fallacy," which is a mistake in ethical reasoning that some naturalists make when they believe that something is good, for instance, just because it comes from nature—or so say ethical philosophers such as G. E. Moore, *Principia Ethica* (Cambridge, England: Cambridge University

Press, 1903); this is the book that coined the phrase "naturalistic fallacy." Essentially, Moore was trying to assert that ethical values—i.e., "what is good?"—cannot be defined by reductionist reasoning. I took a similar route in the text when I stated that "nature" or "natural" is everything and therefore defies any specific definition. But if nature is taken to be equivalent to organic evolution, then it can be analyzed and defined with respect to its parts. This is the contention of many evolutionary biologists, led by those who champion sociobiology. A critic of Moore's reasoning is naturalist E. O. Wilson, who contends that what is good ultimately comes from the natural history of the human species. He would suggest that "good" comes from mental adaptation. A sense of goodness is a property of our brains, which exists within a cultural setting. Since it is part of the gene-culture coevolution of humans, Wilson would say, a sense of what is good is not absolute and static, but rather malleable through time, potentially changing with the conditions of the surrounding environment. See E. O. Wilson, *Consilience, the Unity of Knowledge* (New York: Alfred A. Knopf, 1998), 248–251.

I prefer to sidestep this dichotomous debate by fusing elements from both schools of thought. I am sensitive to Moore's aversion to ethical reductionism, but I also agree with Wilson that human natural history set the stage for our ethical impulses. Philosophers, legislators, and political commentators should be well-versed in evolutionary biology, in my opinion. But like the discussion of atheism, debates on ethical philosophy can derail the important work of solving more important problems in biological science.

If I'm forced to enter into a discussion of ethics, I would say that I don't think evolutionary history has caused us to be inherently good or inherently evil. I recognize that all of human biology comes from what we call "nature," and things like notions of good or bad are amplified or diminished by a set of causes (called "culture") that are superimposed upon those of organic evolution. I prefer to maintain that ethics, and all those topics subsumed under that heading, are imparted in our early years of life by the people who surround us—usually our immediate family members. All of our psychological traits are affected throughout life, of course, but in those early years our deepest sense of right and wrong is probably established.

4. For more information, visit the Society for Ecological Restoration International Web site at http://www.ser.org. Also, the July 31, 2009, issue of *Science* magazine was dedicated to the latest research in restoration ecology.

Learning about sustainable development of natural resources and restorative practices can be achieved even if you have no interest in attending college. Some of the most practical information comes from a federally mandated community outreach program called Agricultural Cooperative Extension, an outgrowth of America's "land grant" universities. Many of our country's best research institutions were established by the Morrill Act in 1862, during Abraham Lincoln's presidency. That same year saw the creation, by Lincoln, of the U.S. Department of Agriculture (USDA).

The Morrill Act established land grant colleges in every state, with the intention of emphasizing agricultural, mechanical, home-economical, and forestry professions at the level of higher education. In 1887, the Hatch Experiment Station Act created a federal fund to support agricultural research, sponsored by the USDA, in the interest of extending practical research from experiment stations to the farmers and citizens in all states.

Today, agriculture and ecological health are intertwined. The research stations and their scientists therefore deal with ecological issues that extend far beyond the local community. Many agricultural practical problems are recognized as having a regional or global scale, such as droughts, erosion, or climate change. The cooperative extension programs, therefore, have become important disseminators of information about stewardship of natural resources, including restoration ecology, which melds nicely with their primary focus on practical agricultural research.

To learn more about the local Cooperative Extension in your community, go to http://www.extension.org and type in your zip code.

5. Whatever the reason, I know that others have felt the same emotional connection: "But few indeed, strong and free with eyes undimmed with care, have gone far enough and lived long enough with the trees to gain anything like a loving conception of their grandeur and significance as manifested in the harmonies of their distribution and varying aspects throughout the seasons, as they stand arrayed in their winter garb rejoicing in storms, putting forth their fresh leaves in the spring while steaming with resiny fragrance, receiving the thunder-showers of summer, or reposing heavy-laden with ripe cones in the rich sungold of autumn. For knowledge of this kind one must dwell with the trees and grow with them, without any reference to time in the almanac sense." From John Muir, *The Mountains of California* (New York: Century Co., 1907), 140.

6. This scenario is based on calendar dates (calibrated radiocarbon dates) reported by Arthur Bloom, "The Late Pleistocene Glacial History and Environments of New York State Mastodons," *Palaeontographica Americana* 61 (2008), 13–24. The glacial maximum in this part of North America was between roughly 23,000 and 26,000 years ago. So I am assuming a time frame of somewhere in the middle of the glacial retreat between the glacial maximum and the "Younger Dryas" period, which is most recently calculated to be 12,900–11,600 years ago. See Michael Balter, "The Tangled Roots of Agriculture," *Science* 327 (2010), 404–406. The descriptions that follow later in the chapter are based on Warren Allmon and Peter Nester, eds., "Mastodon Paleobiology, Taphonomy, and Paleoenvironment in the Late Pleistocene of New York State: Studies on the Hyde Park, Chemung, and North Java Sites," *Paleontographica Americana* 61 (2008),

and P. T. Davis et al., "Quaternary and Geomorphic Processes and Landforms Along a Traverse Across Northern New England, U.S.A.," in D. J. Easterbrook, ed., *Quaternary Geology of the United States, INQUA 2003 Field Guide* (Reno, Nevada: Desert Research Institute, 2003), 365–398.

7. As of this writing, the International Union of Geological Sciences had redefined the base of the Pleistocene based on recommendations from the International Commission on Stratigraphy (http://www.stratigraphy.org). It was decided that, based on new data, the Pleistocene epoch should be extended to include almost 800,000 years of glacial activity, pushing the starting date of the epoch back to roughly 2.5 million years before the present. New data and new committees might change the date again. For the purposes of this book, I shall refer to a rough date of 2 million years to denote the beginning of the Pleistocene.

8. For all the tree species mentioned in the last two paragraphs, extinction is not guaranteed. After all, the only requirement for evolution is continuity of reproduction, and immature specimens can be found throughout the forests. But clearly the age structure of those populations has been significantly altered. It's a safe bet to assume that if parasitism has already destroyed the most successful reproducers (the dominant trees in the stand), and the remaining trees are quickly diseased, as they reach maturity, with increased competition from healthier species and intensified efficiency of the parasites, extinction is a serious possibility. The damage to the tree species from insects and fungi could be total unless immunity can evolve faster than parasite specialization.

9. Warren D. Allmon, Peter L. Nester, and John J. Chiment, "Introduction: New York State as a Locus Classicus for the American Mastodon," *Palaeontographica Americana* 61 (2003), 5–12.

10. See George C. Frison, "Paleoindian Large Mammal Hunters on the Plains of North America," *Proceedings of the National Academy of Sciences, U.S.A.* 95 (1997), 14576–14583, and S. Kathleen Lyons, Felisa A. Smith, and James H. Brown, "Of Mice, Mast-

odons and Men: Human-Mediated Extinctions on Four Continents," *Evolutionary Ecology Research* 6 (2004), 339–358.

11. See C. Vance Haynes Jr., "Younger Dryas 'Black Mats' and the Rancholabrean Termination in North America," *Proceedings of the National Academy of Sciences, U.S.A.* 105 (2008), 6520–6525. This hypothesis is strengthened by archaeological sites at ancient watering holes where congregating mega-mammals, attracted to the scarce resource, were ambushed and slaughtered in large numbers by Clovis hunters. But this hypothesis does not yet have enough evidence to be generally accepted. For example, other researchers have suggested that the only way to provoke such an ecological catastrophe is an extraterrestrial impact, and geochemical signatures in sediments from around the Pleistocene-Holocene transition lend some support to this hypothesis.

12. See Nigel Calder, *Magic Universe, the Oxford Guide to Modern Science* (New York: Oxford University Press, 2003). Recent data on fisheries can be found in Boris Worm et al., "Rebuilding Global Fisheries," *Science* 325 (2009), 578–585.

Chapter 9: A Meaningful Afterlife

1. From Aldo Leopold, *A Sand County Almanac* (New York: Oxford University Press, 1949), last line of preface.

2. C. Wright Mills, *The Sociological Imagination* (New York: Oxford University Press, 1959), 196.

3. Typified by the band, the Eagles: " . . . the Eagles became point men along the [Sunset] Strip for what was called, variously, the Mellow Mafia, the Southern California Mafia, and the Avocado Mafia, a seventies version of Frank Sinatra's fifties Hollywood Rat Pack." From Marc Eliot, *To the Limit, the Untold Story of the Eagles* (New York: Little Brown, 1998), 5. The friend I refer to in this section is Laurie Vitt of the University of Oklahoma.

4. Central Intelligence Agency, *The World Factbook*, Washington, D.C., 2010. See https://www.cia.gov/library/publications/the-world-factbook/rankorder/2127rank.html.

5. Douglas L. T. Rohde, Steve Olson, and Joseph T. Chang, "Modeling the Recent Common Ancestry of All Living Humans," *Nature* 431 (2004), 562–566.

6. If a population were completely isolated from all other human populations, the most recent common genealogical ancestor for all living humans would have to predate the isolation of that population. But a careful examination of human societies, as described in ibid., reveals no such population for which reproductive isolation can be guaranteed for more than a few hundred years. Even North and South America before Columbus have remained genealogically connected by continued migration across the Bering Strait.

7. Duncan J. Watts, *Six Degrees: The Science of a Connected Age* (New York: W. W. Norton, 2003).

8. Judith S. Kleinfeld, "Could It Be a Big World After All?" *Society* 39 (2002), 61–66.

9. Jon M. Kleinberg, "Navigation in a Small World," *Nature* 406 (2000), 845.

10. Albert-László Barabási, *Linked: The New Science of Networks* (New York: Perseus, 2002).

11. The quotation is from the nineteenth-century furniture maker Gustav Stickley. See David Cathers, *Gustav Stickley* (New York: Phaidon, 2003).